Wireless Power Transfer: Foundations, Innovations and New Probes

Wireless Power Transfer: Foundations, Innovations and New Probes

Edited by **Shawn Guy**

LANRYE
INTERNATIONAL

New Jersey

Published by Clanrye International,
55 Van Reypen Street,
Jersey City, NJ 07306, USA
www.clanryeinternational.com

**Wireless Power Transfer: Foundations, Innovations
and New Probes**
Edited by Shawn Guy

© 2015 Clanrye International

International Standard Book Number: 978-1-63240-525-8 (Hardback)

Printed in the United States of America.

Contents

Preface

This book provides innovative insights into wireless power transfer. The theme of this book is engineering technology; it extensively classifies diverse classes of wireless power transfer. This book is a collection of modern studies and advancements in the field of wireless power transfer techniques. It contains several chapters that focus on topics of wireless power links, and several system issues in which systematic methodologies, numerical simulation methods, and applicable examples are evaluated.

All of the data presented henceforth, was collaborated in the wake of recent advancements in the field. The aim of this book is to present the diversified developments from across the globe in a comprehensible manner. The opinions expressed in each chapter belong solely to the contributing authors. Their interpretations of the topics are the integral part of this book, which I have carefully compiled for a better understanding of the readers.

At the end, I would like to thank all those who dedicated their time and efforts for the successful completion of this book. I also wish to convey my gratitude towards my friends and family who supported me at every step.

Editor

Network Methods for Analysis and Design of Resonant Wireless Power Transfer Systems

Marco Dionigi[1], Alessandra Costanzo[2] and Mauro Mongiardo[1]
[1]University of Perugia
[2]University of Bologna
Italy

1. Introduction

1.1 A taxonomy of electromagnetic energy transfer

Efficient energy transfer is nowadays becoming an essential topic: it allows to save economical resources, to improve the quality of life, to reduce pollution, to name just a few issues. In particular, energy transfer is often realized by using electromagnetic power which can either be transmitted along a guiding medium (transmission lines), or without a supporting medium. Waveguiding structures can possess a discrete spectrum, as e.g. metallic waveguides or, in the instance of waveguides with unbounded cross section, a continuous spectrum (or both). These cases are well known and a considerable literature exists for their analysis and design, see e.g. Collin (1960), Rozzi (1997). Unfortunately, guided waves require the presence of a medium to support wave propagation; in addition, energy decays along the transmission lines in an exponential manner.

A different mechanism for transmitting electromagnetic energy is that of radiation; note that also in this case a generalized network representation is feasible in terms of spherical modes, as described in Felsen (2009), Mongiardo et al. (2009). Naturally, at microwave frequencies, the antennas become more directive as the frequency is increased (and, therefore, the dimensions of the antenna with respect to the wavelength are also increased). On the other hand, propagation at very high frequencies may pose problems due to the atmospheric attenuation, availability of hardware components, etc. In addition, radiated energy is spread over radiation angles and its attenuation goes with an inverse quadratic law with respect to distance, even in the case of vacuum.

Another possibility for transmitting electromagnetic energy is by using the reactive fields present nearby open resonators, as recently suggested in Kurs et al. (2007) and Karalis et al. (2008). In fact, two open structures operating at the same resonant frequency may exchange energy. It suffices a very low coupling for the energy transfer to take place and, in presence of "perfects" infinite Q resonators, even a very weak coupling will provide the possibility of transfering energy to the load (which is the only places where energy can be dissipated since, by definition, the resonators are lossless). In practice, by placing many resonators physically separated by one another, it is possible to realize a wireless power transfer based on the near field characteristics and avoiding both radiation and the presence of a conducting media. It is clear that the type of energy transfer that can be achieved depends on the coupling and on the

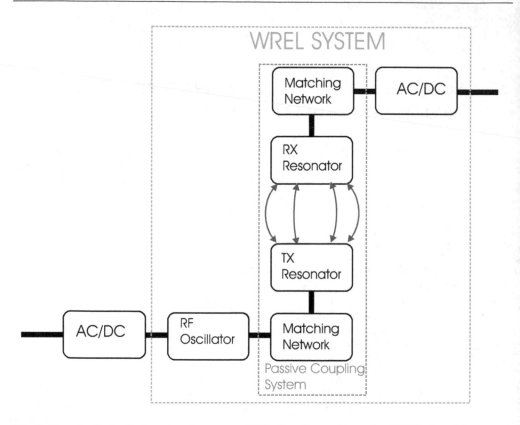

Fig. 1. Sketch of a typical system for resonant Wireless Power Transfer (WPT). Apart for AC/DC converters we note the presence of a RF oscillator and a transmitting (TX) resonator, a receiving (RX) resonator and matching networks to couple the energy from the RF oscillator to the TX resonator and from the RX resonator to the AC/DC converter. Note that the part between the output of the RF oscillator and the input of the AC/DC converter, denoted as passive coupling system in the figure, is a two port network containing only linear, passive, components.

quality factor of the resonators. As we will see in this chapter, this problem can be formalized in an accurate and rigorous manner by using network theory. By using the latter, it is possible to find out to what extent this resonant reactive field coupling can be employed in various applications. Note that the resonant coupling which we are referring to is very different from inductive coupling, which is suitable only for very short range; a nice comparison between these two types of wireless energy transfer has been published in Cannon et al. (2009). Also observe that we look for magnetic field coupling which has the advantage, with respect to electric field coupling, to be rather insensitive to the presence of different dielectrics. It is also worthwhile to consider that, in order to avoid radiation, it is preferable that the electric field storage takes place in a confined region of space, which may be physically realized either by a lumped capacitance or by other suitable structures.

1.1.1 A word about nomenclature
The energy exchange which we are interested to in this chapter is the one that occurs when two, or more, structure resonates at the same frequency and exchange their energy. Thys type of phenomenology has been referred to in the literature with several different names: Wireless Resonant Energy Links (WREL), Wireless Power Transfer (WPT), WITRICITY (WIreless elecTRICITY), Wireless Energy Transfer (WET), etc. Each different name possesses its own advantages and disadvantages; it is only necessary to keep in mind that, in this work, we are referring to *resonant* structures and we are not referring to antennas (which mainly serve to produce a radiated field). Note, in addition, that a radiated field is, in this type of application, particularly undesiderable since it poses increased problems of electromagnetic compatibility.

1.2 Description of a typical system for medium–range resonant power transfer
In Fig.1 we have reported the sketch of a typical system for WPT. Apart from AC/DC converters we note the presence of a RF oscillator and a transmitting (TX) resonator, a receiving (RX) resonator and matching networks to couple the energy from the RF oscillator to the TX resonator and from the RX resonator to the AC/DC converter. Note that the subnetwork between the output of the RF oscillator and the input of the AC/DC converter is a linear system; this part can be described as a two port network containing only linear, passive, components. An example of a practical implementation of this part is reported in Fig.2. In the next section we will show why this type of structure is advantageous for realizing efficient WPT.

1.3 Plan of the chapter
In sec. 2 we introduce a simple network that, nonetheless, allows us to understand resonant WPT. In that section we will find why it is convenient to introduce matching networks for improving the efficiency of wireless energy transfer. In addition, the complete network (including the matching sections), which describe the linear, passive, part of a WPT system, permits derivation of a relationship between the radio frequency (RF) efficiency and the parameters (quality factor Q, coupling) of the resonators. In this way, theorethical limits can be established and the range for WPT and its efficiency are related to the resonators' parametrization (Q and coupling). A couple of examples of implementations are also illustrated in order to verify what is experimentally feasible for a WPT system operating at a fixed distance.

The next section, sec. 3, deals with the analysis method for a more general case: we consider a methodology which allows the analysis when multiple TX and RX resonators are present. In particular, we note that, by adding further resonators, it is possible to extend the range for WPT. In that section also an approach useful for analyzing this type of network with a common circuit simulator (SPICE) is illustrated.

A crucial block of an inductive power link system is the external power driving unit. The design of this part of the system is considered in sec. 4; in this section we introduce a type of oscillator suitable for resonant WPT. When considering inductive links over variable distances, we observe that the resonant frequency of the system shifts. The proposed power driving unit automatically adjust itself, thus providing a convenient implementation. An example, along with its experimental verification, is also presented.

The quality factor of the resonators is also one of the key elements in the design of resonant WPT system; sec. 5 describes how to measure the Q for a given resonator. Finally, conclusions are summarized in the last section.

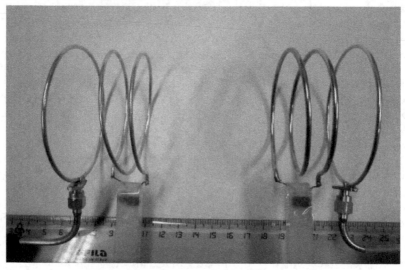

Fig. 2. A structure showing the inductive coupling between the source (first loop on the left), the transmitting resonator (second coil from the left), receiving resonator (third coil from the left) and load (last loop on the right).

2. Network theory for medium–range power transfer

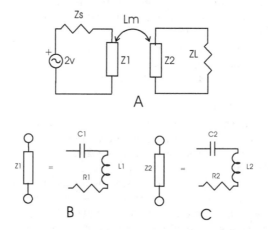

Fig. 3. A simple network for the study of wireless resonant energy links. The two resonators are described by Z_1 and Z_2; R_1 and R_2 take into account the losses in the resonators.

2.1 A simple network model for coupled resonators

A very simple network for understanding WREL is reported in Fig. 3, where Z_1 and Z_2 take into account the resonant element values, and Z_S, Z_L are the source and load impedances. By placing a generator of amplitude 2, the computation of S_{21} is quite straightforward. By denoting with I_1 the current flowing in the circuit on the left side and with I_2 the current flowing in the circuit on the right sight, we may write Kirchhoff's voltage laws as:

$$2 = (Z_S + Z_1)I_1 - j\omega L_m I_2 \tag{1}$$
$$0 = (Z_L + Z_2)I_2 - j\omega L_m I_1 \ . \tag{2}$$

After finding I_1 from I_2 by using the second eq., replacing it in the first eq. and solving, we obtain:

$$S_{21} = Z_L I_2 = \frac{2j\omega L_m Z_L}{(\omega L_m)^2 + (Z_S + Z_1)(Z_L + Z_2)} \tag{3}$$

It is convenient to define the efficiency of the linear, passive, two port network represented in Fig.1 as:

$$\eta = |S_{21}|^2 * 100 \tag{4}$$

By finding I_2 from I_1, and by using the second eq. while denoting with Z_{in} the equivalent impedance in series with Z_S, we have that, for maximum power transfer, the following condition holds:

$$Z_S^* = Z_{in} = Z_1 + \frac{(\omega L_m)^2}{(Z_L + Z_2)} \tag{5}$$

2.1.1 Resonant frequencies

In order to realize resonant WPT, the two resonators must have an equal resonant frequency, i.e. $\omega_R = \frac{1}{\sqrt{L_1 C_1}} = \frac{1}{\sqrt{L_2 C_2}}$; as long as the resonator coupling is weak, the system has only this resonant frequency.

As the coupling increases we need to take into account also the effect of the other resonator. To compute the resonant frequencies, it is advantageous to derive an alternative network. With reference to Fig. 4 it is also possible to note that the circuit represented in Fig. 4A is equivalent to the circuit represented in Fig. 4B.

Let us consider the case of two identical resonators with $L_1 = L_2 = L$ and $C_1 = C_2 = C$. With reference to Fig. 4B, by defining $\omega_0 = \frac{1}{\sqrt{LC}}$ and $k = L_m/L$, and by considering even and odd excitations, corresponding to a magnetic (ω_m) and an electric resonance (ω_e) respectively, we get the following expressions for the resonant frequencies:

$$\omega_m = \frac{1}{\sqrt{(L + L_m)C}} = \frac{\omega_0}{\sqrt{(1 + k)}} \tag{6}$$
$$\omega_e = \frac{1}{\sqrt{(L - L_m)C}} = \frac{\omega_0}{\sqrt{(1 - k)}} . \tag{7}$$

Note that for very small values of k, the resonant frequency is not affected by the presence of the other resonator. Naturally, as the resonators couple their fields, we note that two resonant frequencies appears; in addition, by changing the position of one resonator with respect to

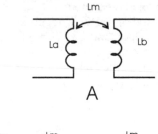

A

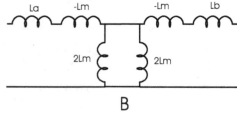

B

Fig. 4. Equivalent network for the coupled inductors. Representation of coupled inductors in terms of impedance inverter.

the other, will shift the resonant frequency. As a consequence, we need to find an oscillator that is able to keep track of these frequency shifts. It is wortwhile to point out that the latter frequency shifts may occur also when inserting more than one receiving resonators.

2.1.2 Maximum efficiency

It is apparent from eq.(3) that the efficiency depends on the values of Z_S and Z_L. In order to analytically compute the maximum efficiency let us consider the case of identical source and load impedance:

$$Z_L = Z_S = Z_0. \tag{8}$$

With reference to the circuit of Fig. 3 we note that the efficiency now depends only on the reference impedance Z_0.

The values of L and C depend on the chosen arrangement (i.e. frequency of operation, dimensions of the coils etc.) and are kept fixed; once the value of k is given, it is possible to compute the efficiency. By considering identical resonators and at resonance

$$Z_1 = Z_2 = R \tag{9}$$

we have, from eq. (3)

$$S_{21}(Z_0) = \frac{2j\omega_0 L_m Z_0}{[\omega_0^2 L_m^2 + R^2 + 2RZ_0 + Z_0^2]}. \tag{10}$$

By taking the derivative w.r.t. Z_0 we have:

$$\frac{\partial \mid S_{21}(Z_0) \mid}{\partial Z_0} = \frac{2j\omega_0 L_m \left(R^2 + \omega_0^2 L_m^2 - Z_0^2\right)}{[\omega_0^2 L_m^2 + R^2 + 2RZ_0 + Z_0^2]^2}. \tag{11}$$

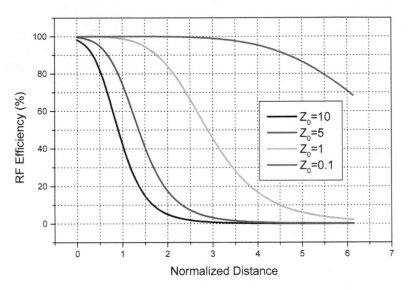

Fig. 5. Computed radio–frequency efficiency (η), for the circuit of Fig. 3, with respect to the normalized distance for various different values of reference impedance Z_0.

The condition

$$\frac{\partial \mid S_{21}(Z_0) \mid}{\partial Z_0} = 0 \tag{12}$$

provides the sought value for the reference impedance:

$$Z_0 = \sqrt{R^2 + \omega_0^2 L_m^2} \tag{13}$$

or equivalently:

$$\omega_0^2 L_m^2 = Z_0^2 - R^2; \tag{14}$$

By inserting the above expression in denominator of eq.(10) we obtain for the S_{21}

$$S_{21} = \frac{j\omega_0 L_m}{[R + Z_0]} \tag{15}$$

which provides:

$$\mid S_{21}(Z_0) \mid^2 = \frac{\omega_0^2 L_m^2}{[Z_0 + R]^2} = \frac{[Z_0 - R]}{[Z_0 + R]} \tag{16}$$

In Fig. 5 we have plotted the radio–frequency efficiency (η of (4)) with respect to the normalized distance for various different values of reference impedance Z_0. It is important to note that high values of efficiency are possible only if the reference impedance is low. This means that, if the distance is fixed, in order to achieve a high efficiency it is appropriate to use matching networks between the generator and the transmitting coil and between the load and the receiving coil, that transform the impedances of the load and generator into suitable values.

2.2 Matching networks

In order to improve the efficiency we have seen that it is appropriate to introduce matching networks between the source and the transmitting resonator and between the receiving resonator and the load. These matching network will adjust the source/load impedance to the impedances necessary to optimize the efficiency. A possible way to realize the matching networks is by inserting impedance inverters before the load impedance and after the source impedance; by doing so we recover the equivalent network shown in Fig. 6.

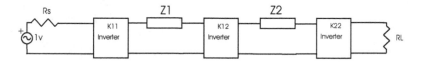

Fig. 6. Narrowband circuit model using impedance inverters; note the presence of the couplings K_{11} and K_{22} to model the input and output loop coupling to the resonator .

Impedance inverters can be realized in a variety of ways. As an example, let us recall the equivalences reported in Fig. 7, Fig. 8, where it is shown that immittance inverters can be represented, over a narrow band, in terms of inductive or capacitive networks.

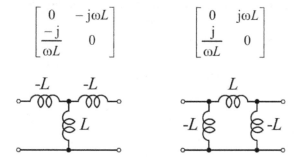

Fig. 7. Realization of immittance inverters in terms of inductive networks. The corresponding ABCD matrices are also reported.

An alternative approach may also be employed, by using the same resonant coils also as matching networks. In fact, with reference to Fig. 9, we note that we have connected the coaxial center line at a certain distance from the end of the coil, thus splitting the inductance. It may be noted that the equivalences reported in Fig. 10 may be used. Therefore, also in this case, the representation in terms of immittance inverters can be used. We can therefore conclude that a fairly general network model for representing WRELs is the one reported in Fig. 6.

One possible physical realization is e.g. that shown in Fig. 2; the latter is composed of four coils and two capacitors in a symmetrical structure. We have used 2 mm gauge silver plated wire to wind the coils and two 330 pF silvered mica capacitors to realize the resonators. The input and output coils are 65mm diameter single loops, while the resonator coils are 2 turn, 60 mm diameter, 22 mm length coils. The resonator and the input loops are placed on plastic stands for their positioning.

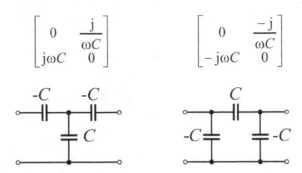

$$\begin{bmatrix} 0 & \dfrac{j}{\omega C} \\ j\omega C & 0 \end{bmatrix} \qquad \begin{bmatrix} 0 & \dfrac{-j}{\omega C} \\ -j\omega C & 0 \end{bmatrix}$$

Fig. 8. Realization of immittance inverters in terms of capacitive networks. The corresponding ABCD matrices are also reported.

It is worthwhile to point out that a rigorous model of the structure in Fig. 2, would have required to take into account also all the couplings between all the resonators. A rigorous network for this situation will be illustrated in sec. 3. For the moment, it suffice to say that the network of Fig. 6 is a very good approximation and applicable for the design of the input/output coupling sections.

Note that inductive or capacitive networks can also be used to shift the resonant frequencies.

Fig. 9. The coaxial input placed along the helix provides a splitting of the inductance that can be used for matching purposes, thus avoiding the use of additional transformers.

2.3 Narrow band analysis of coupled resonators systems

The equivalent network shown in Fig. 6 can be analyzed in a very simple way by using ABCD matrices. We consider coupled resonators identified by their couplings, Q factors, and resonance frequency; we may express the impedance Z_1 and Z_2 in terms of the resonant frequency $\omega_o = \dfrac{1}{\sqrt{LC}}$, unloaded Q factor, and resonator reactance slope parameter $\chi = \sqrt{\dfrac{L}{C}}$, as follows:

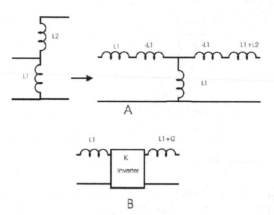

A

B

Fig. 10. Equivalent network showing how a splitted inductance can be represented in terms of immittance inverters.

$$Z_{1,2} = \chi_{1,2} \left[j \left(\frac{\omega}{\omega_o} - \frac{\omega_o}{\omega} \right) + \frac{1}{Q_{1,2}} \right]. \tag{17}$$

The inductors' couplings can be modelled as impedance inverters whose value is:

$$K = \omega L_m = \omega k_{ab} \sqrt{L_a L_b}, \tag{18}$$

where L_m is the mutual inductance and k_{ab} the inductive coupling coefficient, L_a and L_b the inductances of the coupled inductors.

The ABCD matrix of a impedance inverter can be written as follows:

$$\begin{bmatrix} 0 & -jK \\ \frac{-j}{K} & 0 \end{bmatrix}, \tag{19}$$

while for a series impedance Z we have:

$$\begin{bmatrix} 1 & Z \\ 0 & 1 \end{bmatrix}. \tag{20}$$

We can express the input-output ABCD matrix $\mathbf{T}_{ABCD}$ of the cascade of impedance inverters and series impedance of Fig. 6 by the following matrix product:

$$\mathbf{T}_{ABCD} = \begin{bmatrix} A & B \\ C & D \end{bmatrix} = \begin{bmatrix} 1 & Z_s \\ 0 & 1 \end{bmatrix}.$$
$$\begin{bmatrix} 0 & -jK_{11} \\ \frac{-j}{K_{11}} & 0 \end{bmatrix} \cdot \begin{bmatrix} 1 & Z_1 \\ 0 & 1 \end{bmatrix}.$$
$$\begin{bmatrix} 0 & -jK_{12} \\ \frac{-j}{K_{12}} & 0 \end{bmatrix} \cdot \begin{bmatrix} 1 & Z_2 \\ 0 & 1 \end{bmatrix}. \tag{21}$$
$$\begin{bmatrix} 0 & -jK_{22} \\ \frac{-j}{K_{22}} & 0 \end{bmatrix} \cdot \begin{bmatrix} 1 & Z_L \\ 0 & 1 \end{bmatrix}$$

The elements of the ABCD matrix $\mathbf{T}_{ABCD}$ are the following:

$$A = \frac{j\left[Z_2 K_{11}^2 + Z_s K_{12}^2 + Z_1 Z_2 Z_s\right]}{K_{11} K_{12} K_{22}}$$

$$B = \frac{j\left[K_{11}^2 K_{22}^2 + Z_2 Z_L K_{11}^2 + Z_L Z_s K_{12}^2 + Z_1 Z_s K_{22}^2 + Z_1 Z_2 Z_L Z_s\right]}{K_{11} K_{12} K_{22}}$$

$$C = \frac{j\left[K_{12}^2 + Z_1 Z_2\right]}{K_{11} K_{12} K_{22}}$$

$$D = \frac{j\left[Z_L K_{12}^2 + Z_1 K_{22}^2 + Z_1 Z_2 Z_L\right]}{K_{11} K_{12} K_{22}}$$

(22)

By transforming the ABCD matrix to the scattering matrix, and by considering S_{21}, we obtain:

$$|S_{21}|^2 = \left| \frac{2\sqrt{R_s R_L}}{A \cdot R_s + B + C \cdot R_s R_L + D \cdot R_s} \right|^2 .$$

(23)

Equation (23) gives the efficiency of the network when R_s and R_L are respectively the source and load impedance. It is apparent that, from (17) and (23), we can express the power transmission efficiency in terms of the resonators parameters and their couplings.
When considering a symmetrical system, where we have two identical resonator,$(Z_1 = Z_2 = Z)$, and symmetrical couplings ($K_{11} = K_{22} = K$), eq. (21) can be simplified as follows:

$$A = \frac{j\left[K^2 + Z_s K_{12}^2 + Z_{in}^2\right]}{K^2 K_{12}}$$

$$B = \frac{j\left[K^2 + 2 \cdot Z \cdot Z_{in} K^2 + Z_{in}^2 K_{12}^2 + Z^2 \cdot Z_{in}^2\right]}{K^2 K_{12}}$$

$$C = \frac{j\left[K_{12}^2 + Z^2\right]}{K^2 K_{12}}$$

$$D = \frac{j\left[Z_{in} K_{12}^2 + Z K^2 + Z_{in} Z^2\right]}{K^2 K_{12}}$$

(24)

and eq. (23), with input and output reference impedance R, simplifies into:

$$|S_{12}|^2 = \left| \frac{2R}{A \cdot R + B + C \cdot R^2 + D \cdot R} \right|^2 .$$

(25)

2.4 Input and output coupling design

Let us consider the equivalent network of the WREL system shown in Fig. 6; while the coupling k_{12} between the resonators coils is dependent on their distance and dimensions, the input-output couplings must be designed in order to maximize $|S_{21}|^2$. Input and output impedances, denoted respectively by R_S and R_L, are in general different; also loop resonator resistances R_1 and R_2 may be different. Note that at resonance $Z_1 = R_1$ and $Z_2 = R_2$; in this case it is straightforward to trasfer the output impedance at the input port, giving:

$$R_{in} = \frac{K_{11}^2}{R_1 + \frac{K_{12}^2}{R_2 + \frac{K_{22}^2}{R_L}}},$$

(26)

where K_{11} and K_{22} are, respectively, the input and output inverters values; R_1 and R_2 the loop resistance of the transmitting and receiving resonator, R_L the load resistance, and K_{12} the inverter corresponding to the coupling of the resonators coils. In particular, we can consider $K_{11} = K_{22}$; the matching condition at the input port is:

$$R_{in} = R_S.$$

(27)

It follows from (26) and (27) that:

$$K_{11} = K_{22} =$$

$$= \sqrt{\frac{\frac{R_2 R_L - R_1 R_S}{2} + \sqrt{(R_2 R_L - R_1 R_S)^2 + 4 R_L R_S \left(K_{12}^2 - R_1 R_2\right)}}{2}}. \tag{28}$$

Once the value of the input inverter is obtained it can be realized by different implementations, like coupled inductors, tapped inductors or capacitive networks as described before. If we consider a symmetrical structure where we have $L_1 = L_2$, $Z_1 = Z_2$, $R_s = R_L = R$, from eq. (17) the eq. (28) can be furtherly simplified as follows:

$$k_{11} = k_{22} = \sqrt{\frac{R}{L_1 \omega_o} \sqrt{k_{12}^2 + \left(\frac{1}{Q}\right)^2}}. \tag{29}$$

In the above eq. k_{12} is the inductive coupling coefficient between the resonators, Q the quality factor, ω_o the resonance frequency and $k_{11} = k_{22}$ are the coupling coefficient of the input-output loops to the resonators. Once the input-output coupling coefficient is calculated, one can compute the distance between the coils centers as in Dionigi & Mongiardo (2010).

2.5 Computation of the network elements

The equivalent network is quite useful for modeling wireless resonant energy links in different conditions, as reported in Dionigi et al. (2009) and Dionigi et al. (2010). Moreover, the values of the network elements can be derived as described in many references (see for example sec. 4.1 of Finkenzeller (2003)or Grover (2004) for inductance calculations). In particular, the coil inductance, expressed in μH, of total length L_e, diameter D (expressed in meters), and N turns, can be written as:

$$L = \frac{(ND)^2}{(L_e + 0.45D)}. \tag{30}$$

By denoting with R_i the radius of the i-th coil and by H_{12} the distance between the center of resonator 1 and resonator 2, we can compute the magnetic coupling factor K as in Grover (2004):

$$K_{12} = 1.4 \frac{\left(R_1^2 R_2^2\right)}{\sqrt{R_1 R_2} \sqrt{\left(R_1^2 + H_{12}^2\right)^3}}. \tag{31}$$

Finally, the mutual inductance is computed by:

$$M_{12} = K_{12} \sqrt{L_1 L_2}. \tag{32}$$

2.6 Design of a wireless power transfer system at fixed distance

Fig. 11 illustrates the behavior of the coupling factor k_{12} between two resonant coils as function of the center distance, normalized to the coils diameter. Naturally, the coupling decrease with the distance. Once the value of k_{12} is given, assuming $R = 50\Omega$ and by using (29), it is possible to compute the input coupling coefficients k_{11} and k_{22}. Once the latter coupling coefficients are computed, we can calculate the distance from the input loop of the resonator coil center which

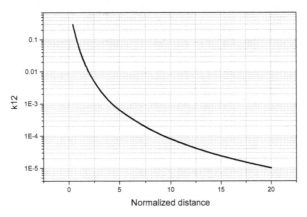

Fig. 11. Coupling coefficient of a 2 turn, 60 mm diameter coil vs center coils distance normalized to diameter.

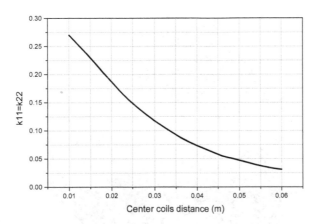

Fig. 12. Input loop and resonator coil coupling coefficient vs center coil distance for 65 mm diameter loop and 2 turn, 60 mm diameter, 22 mm length coil.

realize the sought value. Fig. 12 gives the coupling factor of the input loop to the resonator coil as function of the center distance of the resonator coil.

In Fig. 13 and Fig. 14, the efficiency of the passive, linear part of the system is investigated as function of the distance and resonators Q factor. Note that the proposed design procedure allows, for a given type of resonator, thus for given Q and coupling, to find the maximum efficiency achievable. In addition, as noted in Dionigi & Mongiardo (2011), the design procedure also permits, for a given distance and efficiency, to find the necessary resonator quality factor Q.

As an example, we have designed two study cases by separating the resonators coil centers by 75mm and 100 mm. The calculated input loop inductance is 130 nH. The transmitting/receiving resonators have been designed for operating at a resonant frequency of 14.56 MHz. The measured Q of the resonators is about 300 and the slope parameter is $\chi=33.26$ Ω. From the design formulas we have calculated the input loop to resonator coil distance of

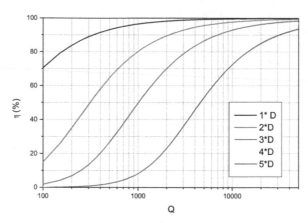

Fig. 13. Computed efficiency for coils diameter D=60mm at distances of 1D-5D, as function of Q factor.

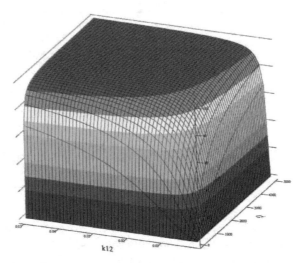

Fig. 14. Computed efficiency as a function of the Q and of the coupling.

15 mm and 25mm, respectively. The scattering parameters of the structure has been measured by a vector network analyzer (agilent PNA n5230A) assuming input and output impedance $R = 50\Omega$. In Fig. 15 are reported the calculated and measured efficiency of the coupled resonator system; it is possible to observe a fairly good agreement. It can be also noted that, at 100mm distance, it is not possible to reach very high efficiency values; this is mainly due to the low Q factor of the resonators. It is clear from the theory that higher transmission efficiencies can be reached only if very high Q resonators are used (see also Dionigi & Mongiardo (2011)). As a further example, the structure of Fig. 16 has been optimized for WPT at different distances. The system is composed of a pair of resonator made of a single loop inductor of diameter 20 cm and a couple of capacitors made of two rectangular pieces of high quality teflon substrate metallized on both sides. In this way a high quality capacitor is obtained.

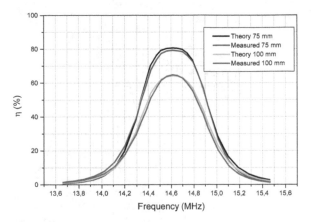

Fig. 15. Simulated and measured efficiency at 75 and 100 mm resonators coils center distance.

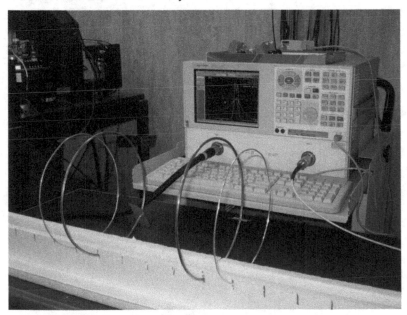

Fig. 16. A resonant WPT link with matching networks.

The teflon substrate is a taconic TLY5 with thickness of 0.51mm. A capacitance of about 72 pF is easily obtained, and, by trimming the edge of the substrete, an accurate tuning of the resonators is obtained. The system has been tuned and measured. In Fig. 17 -Fig. 18 and 19 we have reported the efficiencies of the linear part as obtained by using a Vector network analizer.

Due to the resonators quality factor of about 350 a high efficiency is obtained at a distance of 10 cm. In this configuration the linear part of the WPT system delivers the power to the load almost like a direct connection. As the distance between the two resonators increases a drop

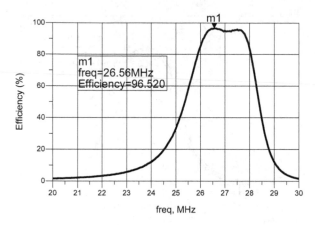

Fig. 17. Measured efficiency at 10 cm resonators coils center distance; the corresponding normalized distance is 0.5. The normalized distance is defined as the ratio between the coils distance and their diameter.

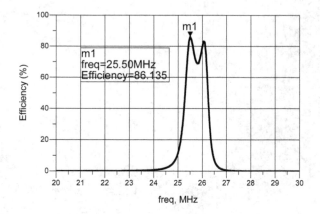

Fig. 18. Measured efficiency at 20 cm resonators coils center distance; the corresponding normalized distance is 1.

of the efficiency is measured, although at the distance of 30 cm an efficiency of about 77% is reached.

3. General analysis of WRELs

3.1 The general network model for coupled resonators
A more general analisys technique can be adopted when multiple resonators are employed. In this case, cross coupling between resonators is possible and a more accurate analisys can be implemented. In Fig. 20, as an example, three mutually coupled inductors are shown. An equivalent network of the structure of Fig. 20 is illustrated in Fig. 21.

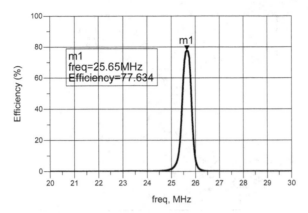

Fig. 19. Measured efficiency at 30 cm resonators coils center distance; the corresponding normalized distance is 1.5.

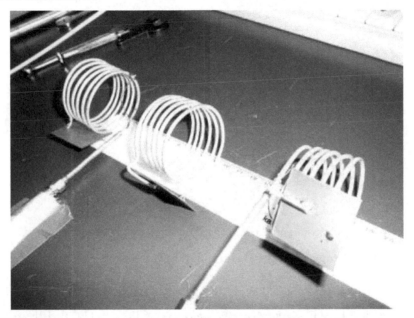

Fig. 20. Photograph of resonant conducting wire loops: the first and last resonator are, respectively, the source and the load. The resonator in the middle extends the operating range.

We can compute the solution of the circuit of Fig. 21 in the standard manner. By defining the impedance Z_i with $i = 1, 2, 3$ as

$$Z_i = j \left(\omega L_i - \frac{1}{\omega C_i} \right) + R_i \tag{33}$$

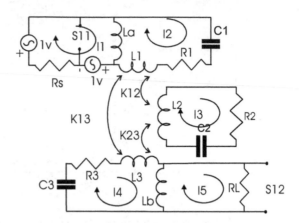

Fig. 21. Equivalent network of the wireless resonant energy link configuration shown in the previous figure.

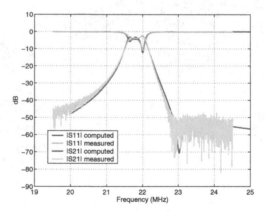

Fig. 22. Reflection and transmission coefficient of the circuit with three resonators.

and by calling with A the system matrix and with I the currents vector we can write:

$$A \cdot I = \begin{bmatrix} j\omega L_a + R_s & -j\omega L_a & 0 & 0 & 0 \\ -j\omega L_a & Z_1 & j\omega M_{12} & j\omega M_{13} & 0 \\ 0 & j\omega M_{12} & Z_2 & j\omega M_{23} & 0 \\ 0 & j\omega M_{13} & j\omega M_{23} & jZ_3 & -j\omega L_b \\ 0 & 0 & 0 & -j\omega L_b & j\omega L_b + R_L \end{bmatrix} \begin{bmatrix} I_1 \\ I_2 \\ I_3 \\ I_4 \\ I_5 \end{bmatrix} = \begin{bmatrix} 2 \\ 0 \\ 0 \\ 0 \\ 0 \end{bmatrix} \qquad (34)$$

For scattering parameters computation it is possible to use the excitation scheme shown in Fig. 21 with two 1 V generators; by considering this arrangement we have:

$$S_{11} = 1 + I_1 R_s; S_{21} = -I_5 R_L \qquad (35)$$

The measured and calculated response of the circuit of Fig 21 is shown in Fig. 22. It is apparent that a direct coupling between the input resonator and the output one adds a transmission zero to the transmission coefficient.

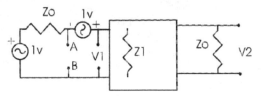

Fig. 23. Equivalent network for spice computation of scattering parameters.

The analysis of a multicoil system can be obtained extending eq.s (34-35). In the general case no assumption can be made on the mutual couplings of the different coils. This affects the solution matrix producing non zero elements above and below the second diagonal row. Discarding the capacitive coupling the solution matrix elements can be written as follows:

$$\mathbf{A}_{i,k} = \begin{cases} -j\omega M_{i,k} & \text{if } i \neq k \\ j\left[\omega L_i - \frac{1}{\omega C_i}\right] + R_i. & \text{if } i = k \end{cases} \tag{36}$$

This methodology, although well known, pave the way to the analysis and optimization of very complex structures that may arise from the necessity of extending the range of a WPT system while controlling the field values.

3.2 Spice computation of scattering parameters
With reference to the circuit illustrated in Fig. 21, our goal is to obtain a characterization of the circuit. A simple modification of the input section allows us to find out the scattering parameters directly from spice computations. By considering Fig. 23 and by denoting with Z_0 the reference impedance, we have

$$S_{11} = \frac{Z_1 - Z_0}{Z_1 + Z_0} \tag{37}$$

and

$$V_1 = 2\frac{Z_1}{Z_1 + Z_0}. \tag{38}$$

By solving the above eq. for Z_1

$$Z_1 = 2\frac{V_1 Z_0}{2 - V_1}, \tag{39}$$

and, by substituting into (37), we obtain:

$$S_{11} = V_1 - 1 \tag{40}$$

From the circuit in Fig. 23 it is apparent that the voltage between points A and B provides the scattering parameter S_{11}.
Concerning S_{21} we have:

$$S_{21} = \frac{V_2 - I_2 Z_0}{V_1 + I_1 Z_0}. \tag{41}$$

When the port 2 is terminated by its reference impedance the following relationship holds:

$$I_2 = -\frac{V_2}{Z_0}, \tag{42}$$

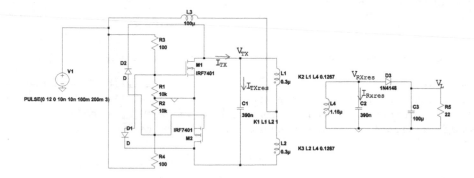

Fig. 24. Power driving unit schematic, based on a Royer oscillator.

which yields:

$$S_{21} = 2\frac{V_2}{V_1 + I_1 Z_0}. \tag{43}$$

By considering that:

$$I_1 = \frac{V_1}{Z_1} \tag{44}$$

we get:

$$S_{21} = 2\frac{V_2}{V_1}\frac{1}{1 + Z_0/Z_1} \tag{45}$$

which, after substitution of (39) into (44), provides the sought result:

$$S_{21} = V_2 \tag{46}$$

By using the above formulation it is possible to obtain an accurate frequency domain and time domain analisys of the circuit by using a standard spice symulator.

4. Design of the power driving unit of the wireless link

A crucial block of an inductive power link system is the external power driving unit. Usually, class E power amplifiers are employed; unfortunately, they require square pulses to drive the circuit which, in turn, may affect the optimum value of the power added efficiency (PAE). A power oscillator is therefore a better choice and it is designed to simultaneously ensure high efficiency and higher harmonics rejection. A possible equivalent circuit topology is schematically reported in Fig. 24. It consists of a differential power oscillator formed by combining a cross-coupled MOSFET structure with a resonant load network. The latter is used as the primary side link coil of the power link system. For power transfer a secondary resonator is used, and is also reported in Fig. 24. The entire link of Fig. 24 is considered for the oscillator analysis to account for the variation of the coupling strength with coils distance

D(mm)	K12	f_{osc}(KHz)	P_{DC} (W)	V_{TX}(V)	I_{TXres}(A)	V_{RXres} (V)	I_{RXres} (A)	V_L (V)	P_L (W)
50	0.3426	201.50	53.70	36.80	20.00	34.50	18.30	27.00	33.20
100	0.1267	225.00	43.50	36.80	20.40	28.60	18.44	23.10	24.30
150	0.0603	230.30	19.60	36.70	21.60	16.50	9.40	12.40	7.06
200	0.0334	230.80	12.00	36.50	21.00	9.50	5.40	6.80	2.13
300	0.0133	231.00	9.00	36.70	21.00	4.40	2.35	2.60	0.31

Table 1. Circuit simulated currents voltages and power.

when determining the oscillator actual load. In this way its nonlinear operating regime and oscillation frequency may be computed. At the oscillator side, the source resonator is represented by a resonant circuit (L1,L2,C1). This circuit couples with the secondary resonator (L4,C2) via the magnetic field in free–space (the electric field being mainly confined in the lumped capacity). Thus, the coupling behaviour is described in terms of the inductances L1, L2 and L4 and their mutual coupling K2 and K3.

A circuit level non linear analysis of the oscillator regime may carried out by time-domain simulation or by means of the Harmonic Balance technique, specialized for autonomous circuits Rizzoli (1994). In the latter case, harmonic content of the oscillator waveforms and the actual frequency dependence on the linear subnetwork are accounted for with no limitations. Oscillator steady state regimes for varying distances between primary and secondary coils are then computed together with the associated output power and conversion efficiency. The linear embedding network is either described by rigorous EM analysis or by the equivalent circuit representation previously discussed in this chapter, at all the frequencies of interest, up to the harmonic order needed to accurately describe the oscillator nonlinear regime. For the case study considered here, the primary coil inductance L_p, expressed in μH, with a total length L_e, diameter D (expressed in meters), and N turns, may be modelled by using eq.(30). By denoting with r_i the radius of the i-th coil and by d_{12} the distance between the center of the primary and secondary resonators, the magnetic coupling factor K may be predicted using eq.(31).

The Q factor of the inductor is included in its circuit model to predict the effective load network. The most significant quantities resulting from simulations of the entire power transmission system, are summarized in Table 1 for varying distances between coils; the corresponding coupling factor K_{12}, is also reported in the same table for clarity. For the present analysis the supply voltage is fixed at 12V and a reference system load of 22 Ohm is considered. From the computed DC current (I_{DC}), the oscillator power consumption (P_{DC}) is derived. The phasors of the output voltages and currents, V_{TX} and I_{TX}, at oscillation frequency, are used to predict the transmitter output power, the power entering transferred to the oscillator load. As expected, V_{TX} is proportional to the supply voltage and does not significantly vary with the coil distances and thus with the system load. For completeness currents and voltages in the secondary resonator (I_{TXres}, I_{RXres}) and the current in the primary one I_{TXres}, are reported in the same table too.

The system dependence on the coils distances and on the coupling factor is evident from the voltages and currents phasors of the receiver resonator branches, V_{RX} and I_{RX}, and of the load branch, V_L and I_L. By means of these quantities the system efficiency, computed as:

$$\eta_{osc}(\%) = \frac{P_L}{P_{DC}} \qquad (47)$$

may be quantified and plotted as in Fig. 26. It can be observed that the transmission efficiency is a nonlinear function of the transmitter and receiver distance and can only be predicted by the simultaneous co-design of the power oscillator and of the coupling resonators. For example, it is interesting to note that transmission efficiency depends on the intensity of the H field. This dependence tells us that, in order to have high efficiency, we have to position the receiver in the high H field intensity zone of the transmitter. Once again, the current flowing in the transmitting resonant circuit, I_{TXres}, is also used to evaluate the magnetic field produced by the transmitting coil in the axial position. In particular, we can calculate the value of the magnetic field along the axis at the center of the coil by using the following expression:

$$H = \frac{I_{TXres} N r^2}{(2(r^2 + d^2))^{3/2}} \tag{48}$$

where I_{TXres} is the current in the coil, N is the number of turns, r the radius of the coil and d the distance of the observation point along the coil axsis fron the center of the coil. Measured and computed values of the magnetic field are compared in Fig. 27.

Previous results may be used as a suitable starting point for a dedicated system design in a specific application environment. For example, once the range of primary and secondary coil distances is established, the desired oscillator band is derived and a broadband optimization may be carried out to enhance the starting point performances in terms of output power and conversion efficiency. These quantities directly determine the efficiency of the entire wireless power transmission system. Design variables may be the DC voltage range and the oscillator linear embedding network, including geometrical parameters of the primary and secondary coils.

For the setup shown in Fig. 25, the resonant coils are made of a 2 turn silver-plated wire with 3 mm gauge wounded with 180 mm diameter, the coils length are about 30 mm. For the starting point system analysis, both coils are connected in parallel with a 390 pF polypropylene capacitor. Their lengths may be adjusted to tune the resonant frequencies.

5. Resonators' quality factor measurement

5.1 Measuring resonant frequency and Q factor

The measurement of the resonant frequency and Q factor of resonators is based principally on two techniques, namely the reflection method and the transmission method (see e.g. pag. 53 and following of Kajfez (1998)). It is interesting to note that it is not possible to measure directly the unloaded Q of a resonator but only its loaded Q. This is due to the necessity to couple the resonator to an external circuitry in order to pick the measurement signal. However, a simple modeling of the test structure allows one to de-embed the unloaded Q from the measured loaded Q.

In Fig. 28 an equivalent circuit of a resonator, composed by L_2, C_2 and R_2, coupled to a measuring probe, is described. The measuring probe construction depends on the characteristics of the resonator. In order to measure the Q of a resonator for wireless power transfer, a simple inductive loop is sufficient.

By introducing $\omega_0 = \frac{1}{\sqrt{L_2 C_2}}$ and $Q_0 = \frac{\omega_0 L_2}{R_2}$, the input impedance Z_i at the probe port is given by:

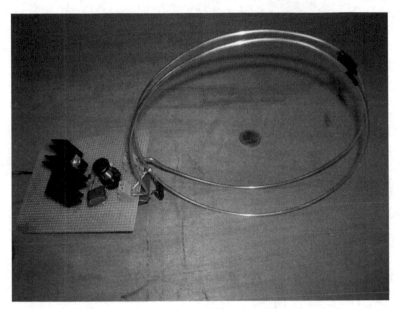

Fig. 25. Photograph of the transmitting circuit and the resonant conducting wire loops together with one euro coin. Note that the resonant coil is excited at its center.

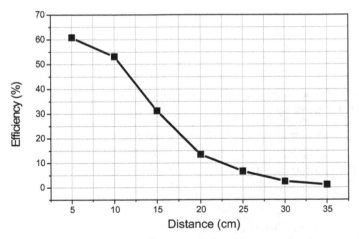

Fig. 26. Computed power transmission efficiency defined as the ratio of the power delivered to the 22 Ohm resistor and the power drawn from the 12 V power supply.

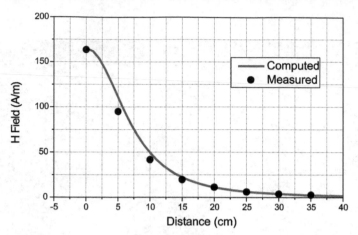

Fig. 27. Measured and computed values of the magnetic field long the axis at the center of the coil.

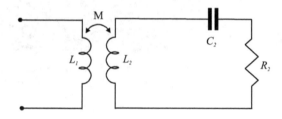

Fig. 28. Equivalent circuit of a resonator (on the right side) coupled to a measuring probe (left side).

$$Z_i = j\omega L_1 + \frac{\frac{(\omega M)^2}{R_2}}{1 + jQ_0 \left(\frac{\omega}{\omega_0} - \frac{\omega_0}{\omega} \right)}. \tag{49}$$

By choosing a small reactance value ωL_1, we can neglect it, and, at the resonant frequency, the input impedance is given by:

$$Z_i = R_i = \frac{(\omega M)^2}{R_2}. \tag{50}$$

It is convenient to further simplify the frequency dependance as follows:

$$\frac{\omega}{\omega_0} - \frac{\omega_0}{\omega} \simeq 2 \frac{\omega - \omega_0}{\omega_0}; \tag{51}$$

accordingly, eq. (49) can be rewritten as follows:

$$Z_i = \frac{R_i}{1 + jQ_0 2 \frac{\omega - \omega_0}{\omega_0}} \tag{52}$$

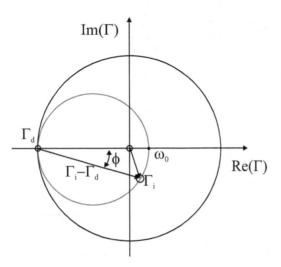

Fig. 29. Reflection coefficient versus ω.

with he corresponding input reflection coefficient expressed as:

$$\Gamma_i = \frac{Z_i - R_c}{Z_i + R_c} \tag{53}$$

where R_c is the reference impedance of the probe port. When we consider a resonant frequency $\omega \to \infty$ we detune the resonator and the input impedance become $Z_i = 0$; the reflection coefficient Γ_d for the detuned resonator becomes $\Gamma_d = -1$.
If we consider the complex number $\Gamma_i - \Gamma_d$, it can be expressed by the following eq.:

$$\Gamma_i - \Gamma_d = \frac{2\frac{R_i}{R_c}}{1 + \frac{R_i}{R_c} + jQ_02\frac{\omega - \omega_0}{\omega_0}}. \tag{54}$$

Given $\kappa = \frac{R_i}{R_c}$ it is clear that it represents the ratio of the power dissipated in the internal resistance and that trasmitted to the output port; by denoting with Γ_{ir} the reflection coefficient at resonance, we have:

$$\Gamma_{ir} - \Gamma_d = \frac{2\kappa}{1 + \kappa} \tag{55}$$

which is a real number. A graphical representation of $\Gamma_i - \Gamma_d$ is illustrated in Fig. 29.
As shown in Kajfez (1998), the following relation holds for loaded and unloaded Q:

$$Q_L = \frac{Q_0}{1 + \kappa} \tag{56}$$

and eq.(54) can be rewritten as follows:

$$\Gamma_i - \Gamma_d = \frac{2}{\left(1 + \frac{1}{\kappa}\right)\left(1 + jQ_L2\frac{\omega - \omega_0}{\omega_0}\right)}. \tag{57}$$

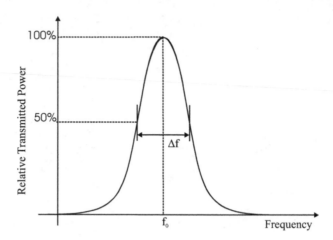

Fig. 30. An example of a measured transmission coefficient.

If we consider the angle ϕ in Fig. 29, we can compute the loaded quality factor Q_L. It can be observed that:

$$\tan(\phi) = -Q_L 2 \frac{\omega - \omega_0}{\omega_0} \tag{58}$$

In order to measure Q_L, one may select two frequencies, denoted by f_3 and f_4, where $\phi = -45°$ and $\phi = 45°$, respectively, thus obtaining:

$$Q_L = \frac{f_0}{f_3 - f_4} \tag{59}$$

Finally, once κ is computed from eq. (55), we have the value of the unloaded Q:

$$Q_0 = Q_L(1 + \kappa). \tag{60}$$

5.1.1 The two port method

The two port measurement can be easily obtained from the previous formulation by considering, instead of the reflection coefficient, the transmission coefficient between two coupled probes. An example of a measured transmission coefficient is given in Fig. 30.
In this case, the relation between the loaded and unloaded Q is the following Kajfez (1998):

$$Q_L = \frac{Q_0}{1 + \kappa_1 + \kappa_2} \tag{61}$$

where κ_1 and κ_2 depend on the input and output coupling.
It is possible to measure the loaded Q factor from the relative transmitted power by simply computing the ratio between the resonant frequency and the half relative power bandwidth as follows:

$$Q_L = \frac{f_0}{\Delta f}. \tag{62}$$

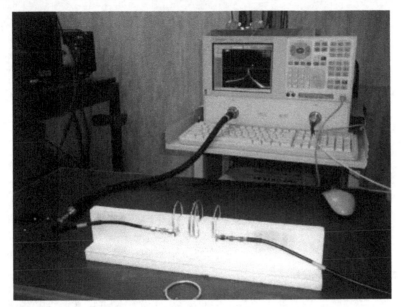

Fig. 31. Measurement setup with inductive loops and the resonator at the center.

Provided that a measure of the values of κ_1 and κ_2 has been performed, the unloaded quality factor is obtained as follows:

$$Q_0 = Q_L(1 + \kappa_1 + \kappa_2). \tag{63}$$

In particular, when the resonator is weakly coupled to the input and output loops, a simpler correction can be applied to the measured loaded Q_L factor as follows:

$$Q_0 = \frac{Q_L}{1 - |S_{21}|^2}. \tag{64}$$

This correction is widely used in practical cases.

5.2 Measurement set-up

In order to measure the resonant frequency and the Q factor of the resonators, two different setups can be adopted which depend on the chosen measurement type. The main instrument needed is a Vector Nework Analyzer (VNA) and a single or a couple of coupling probes. In case of magnetically coupled resonators, circular inductive loops (balanced input) are used; a picture showing the measuring setup is given in Fig. 31.

There are some precautions that have to be followed: the first is to avoid conductive or magnetic materials close to the resonator because they can alter the magnetic field distribution and lead to unaccurate measurement. A second precaution is to test whether is necessary to add a 1:1 balun, between the balanced loop and the coaxial probe, in order to avoid unwanted current distributions on the measuring cables.

The distances between the loops and the resonator have to be adjusted in order to have the smallest coupling which still provide accurate measurements and, usually, an averaged measurement is employed.

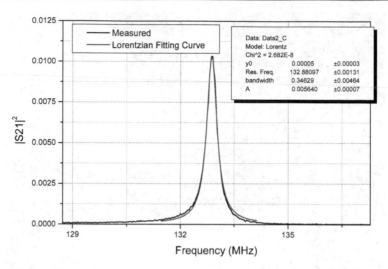

Fig. 32. Lorentzian fitting of the power transmission coefficient.

Once the measurement has been performed, it is possible to obtain the resonant frequency and the loaded Q factor either directly on the VNA, or by performing a fitting of the power transmission coefficient with a suitable peak fitting function.

As an example, a Lorentz function (see Fig. 32), whose expression is given by the following formula, may be used:

$$y(f) = y_0 + \frac{2A}{\pi} \frac{w}{4(f - f_0)^2 + w^2} \tag{65}$$

where y_0 is a constant value A is an amplitude scaling value, w the half power bandwidth and f_0 the resonant frequency. By following this procedure, a quite accurate measurement of resonant frequency and quality factor can be accomplished.

6. Conclusions

We have considered the problem of *resonant wireless electromagnetic energy transfer* over a medium range (from a fraction to a few times the resonator dimension). It has been shown that, by using network theory, considerable progress is feasible and a very high efficiency is attainable for the resonant inductive link. In particular, if the distance between the resonators is kept constant, by a suitable design of the matching networks, an optimal connection can be established for a given quality factor and coupling of the resonators.

For the case of resonators operating over variable distances, a new power oscillator, based on a Royer topology, has been proposed and an entire system has been studied and experimentally verified. A methodology for the analysis of more complex structures, including multiple transmitting and receiving resonators, has also been illustrated and Q measurement has been discussed.

As a final remark we note that resonant wireless power transfer is a new methodology, still under rapid development (see e.g. the notes of the workshop organized by Chen & Russer (2011)), which can be of considerable help in many practical problems: from electrical vehicle charging Imura et al. (2009), to implanted device in the human body Zhang et al. (2009), to immortal sensors Watfa et al. (2010), to battery free operated devices.

7. References

R. E. Collin, (1960). *Field theory of guided waves*, Mc-Graw-Hill Book Co., New York.

T. Rozzi and M. Mongiardo, (1997). *Open Electromagnetic Waveguides*, IEE, London.

L.B. Felsen and M. Mongiardo and P. Russer, (2009). *Electromagnetic Field Computation by Network Methods*, Springer, Berlin.

M. Mongiardo, C. Tomassoni, P. Russer, R. Sorrentino (2009), Rigorous Computer-Aided Design of Spherical Dielectric Resonators for Wireless Non-Radiative Energy Transfer, *MTT-S International Microwave Symposium*, Boston, USA.

A. Kurs, A. Karalis, R. Moffatt, J. D. Joannopoulos, P. Fisher, and M. Soljacic (2007), Wireless Power Transfer via Strongly Coupled Magnetic Resonances, *Science*, 317, pp. 83-86.

A. Karalis, J.D. Joannopoulos, M. Soljacic (2008), Efficient wireless non-radiative mid- range energy transfer," *Annals of Physics*, Elsevier, 323, pp. 24-48.

B.L. Cannon, J.F. Hoburg,D.D. Stancil, S.C. Goldstein (2009), Magnetic Resonant Coupling As a Potential Means for Wireless Power Transfer to Multiple Small Receivers, *IEEE Transactions on Power Electronics* , vol.24, no.7, pp.1819-1825.

M. Dionigi, M. Mongiardo, R. Sorrentino and C. Tomassoni (2009), Networks Methods for Wireless Resonant Energy Links (WREL) Computations, *ICEAA*, Turin, Italy.

M. Dionigi, M. Mongiardo, (2010), CAD of Wireless Resonant Energy Links (WREL) Realized by Coils, *MTT-S International Microwave Symposium*, Anaheim, CA , USA, pp. 1760 - 1763.

M. Dionigi, P. Mezzanotte, M. Mongiardo (2010), Computational Modeling of RF Wireless Resonant Energy Links (WREL) Coils-based Systems , *ACES Conference*, Tampere, Finland.

M. Dionigi, M. Mongiardo, (2011), CAD of Efficient Wireless Power Transmission Systems, *MTT-S International Microwave Symposium*, Baltimore, MD, USA.

M. Dionigi, M. Mongiardo, (2011), Efficiency Investigations for Wireless Resonant Energy Links Realized with Resonant Inductive Coils, *GEMIC, German Microwave Conference*.

K. Finkenzeller (2003), *RFID handbook: fundamentals and applications in contactless smart cards*. Library of congress cataloging in publication data.

W. Grover (2004), *Inductance Calculations*, Dover.

V. Rizzoli, A. Costanzo, F. Mastri and C. Cecchetti (1994), Harmonic-balance optimization of microwave oscillators for electrical performance, steady-state stability, and near-carrier phase noise, *IEEE MTT-S Int. Microwave Symp. Digest*, San Diego, pp. 1401-1404.

D. Kajfez and P. Guillon , (1998). *Dielectric Resonators*, Noble Publishing Corporation, Atlanta.

Z. Chen, P. Russer (organizers), (2011), WFA Workshop on Wireless Power Transmission, *MTT-S International Microwave Symposium*, Baltimore, MD, USA.

T. Imura, H. Okabe, Y. Hori (2009), Basic Experimental Study on Helical Antennas of Wireless Power Transfer for Electric Vehicles by using Magnetic Resonant Couplings, *Vehicle Power and Propulsion Conference, IEEE*, Pages 936-940.

F. Zhang, X. Liu, S.A. Hackworth, R.J. Sclabassi, M. Sun (2009), In vitro and in vivo studies on wireless powering of medical sensors and implantable devices, *Life Science Systems and Applications Workshop, IEEE/NIH*, Bethesda, MD, pp. 84 - 87.

M.K. Watfa, H. Al-Hassanieh, S. Salmen (2008), The Road to Immortal Sensor Nodes, *ISSNIP, International Conference on Intelligent Sensors, Sensor Networks and Information Processing,* Sydney, NSW, pp. 523 - 528.

2

Analysis of Wireless Power Transfer by Coupled Mode Theory (CMT) and Practical Considerations to Increase Power Transfer Efficiency

Alexey Bodrov and Seung-Ki Sul
IEEE Fellow,
Republic of Korea

1. Introduction

In this chapter the analytical model of a wireless power transfer scheme is developed through the means of Coupled Mode Theory (CMT). The derivation is made under the assumption of low internal coil losses and some particular type of resonator (coil inductance and capacitance) equivalent circuit.

With the equivalent circuit modeling the wireless power transfer system the direct high frequency power source connection to the source coil and the usage of external capacitance are considered. It is shown that the maximum efficiency to resonant frequency ratio could be obtained by serial to short end antenna connection of external capacitance.

At MHz frequencies especially near and at the resonant frequencies, the calculation of coil parameters which are the coefficients of the model obtained by means of CMT is not a trivial task. Equivalent resistance, capacitance and inductance of antenna become frequency dependent, and those should be specially considered. Because the equivalent resistance is a critical parameter for the efficiency maximization, the skin and proximity effects are included and the verification of the calculation process is presented. Also due to frequency dependence of equivalent inductance and capacitance, the procedure to obtain the optimal resonant frequency of antennas in terms of the efficiency of the power transfer is discussed.

2. Theoretical analysis of wireless power transfer scheme

In figure 1 a typical diagram of wireless power transfer system is shown. In this system, the inductive reactance and the capacitive reactance of each coil has equal magnitude at the resonant frequency, causing energy to oscillate between the magnetic field of the inductor and the electric field of the capacitor (considering both internal and external capacitances of the coil). The energy transmission occurs due to intersection of magnetic field of the source coil and the load coil. There is no intersection of electric fields, because all electrical energy concentrates in the capacitor (it could be easily shown through Gauss' law of flux).

There are many ways to analyze the wireless power system but here the scattering matrix approach and CMT will be discussed.

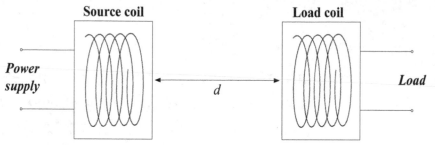

Fig. 1. Typical diagram of wireless power transfer system.

2.1 Efficiency calculation of the wireless power transfer system with scattering matrix's parameters

The wireless power transfer scheme could be analyzed with the two-port network theory, which is formulated in figure 2. As discussed in [1], such networks could be characterized by various equivalent circuit parameters, such as transfer matrix, impedance matrix in (1) and scattering matrix in (2).

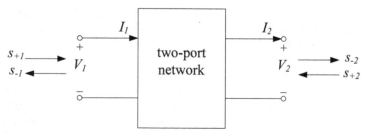

Fig. 2. Two-port network scheme.

$$\begin{bmatrix} V_1 \\ I_1 \end{bmatrix} = \underbrace{\begin{bmatrix} A & B \\ C & D \end{bmatrix}}_{\text{transfer matrix}} \begin{bmatrix} V_2 \\ I_2 \end{bmatrix} \qquad \begin{bmatrix} V_1 \\ V_2 \end{bmatrix} = \underbrace{\begin{bmatrix} Z_{11} & Z_{12} \\ Z_{21} & Z_{22} \end{bmatrix}}_{\text{impedance matrix}} \begin{bmatrix} I_1 \\ -I_2 \end{bmatrix} \qquad (1)$$

where V_1 and V_2 are the input and output voltages of the network and similarly I_1 and I_2 are the input and output currents with the direction specified as in a figure 2. Scattering matrix relates the ingoing $(s_{+1,2})$ and the outgoing waves $(s_{-1,2})$ of the network.

$$\begin{bmatrix} s_{-1} \\ s_{-2} \end{bmatrix} = \underbrace{\begin{bmatrix} S_{11} & S_{12} \\ S_{21} & S_{22} \end{bmatrix}}_{\text{scatering matrix}} \begin{bmatrix} s_{+1} \\ s_{+2} \end{bmatrix} \qquad (2)$$

In electric circuit analysis, transfer and impedance matrices are widely used, but the measurement of coefficients becomes difficult at higher frequencies. Instead, a scattering matrix is preferred due to the existence of network analyzers, which can measure scattering matrix parameters over a wide range of frequencies.

Employing this two-port network concept, the efficiency of power transfer between the generator and the load can be calculated as followings [1].

Analysis of Wireless Power Transfer by Coupled Mode Theory (CMT) and Practical Considerations to Increase
Power Transfer Efficiency

33

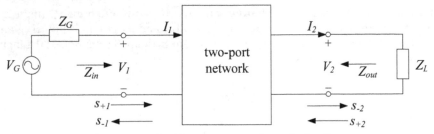

Fig. 3. Two port network connected to the power supply and a load.

From the scattering matrix analysis, the expression for the voltages and currents in terms of wave variables can be presented as (3).

$$V_1 = \sqrt{Z_0}\left(s_{+1} + s_{-1}\right) \quad V_2 = \sqrt{Z_0}\left(s_{+2} + s_{-2}\right)$$
$$I_1 = \frac{1}{\sqrt{Z_0}}\left(s_{+1} - s_{-1}\right) \quad I_2 = \frac{1}{\sqrt{Z_0}}\left(s_{+2} - s_{-2}\right) \tag{3}$$

where Z_0 is the reference impedance value (normally chosen to be 50Ω). Considering figure 3 and (3) it is possible to define scattering matrix equations as (4).

$$\begin{array}{ll} V_1 = Z_{in}I_1 & s_{-1} = \Gamma_{in}s_{+1} \\ V_2 = Z_L I_2 & s_{-2} = \Gamma_L s_{+2} \end{array} \tag{4}$$

where Z_{in} is the input network impedance and Γ_{in}, Γ_L are the reflection coefficients given by (5).

$$\Gamma_{in} = \frac{Z_{in} - Z_0}{Z_{in} + Z_0}$$
$$\Gamma_L = \frac{Z_L - Z_0}{Z_L + Z_0} \tag{5}$$

From (3)-(5) it is possible to define reflection coefficients in terms of scattering matrix parameters.

$$\Gamma_{in} = S_{11} + \frac{S_{12}S_{21}\Gamma_L}{1 - S_{22}\Gamma_L} \tag{6}$$

Following the procedure in Ref.[1], if the roles of the generator and the load are reversed, two more reflection coefficients can be derived as (7)

$$\Gamma_{out} = \frac{Z_{out} - Z_0}{Z_{out} + Z_0} \quad \Gamma_G = \frac{Z_G - Z_0}{Z_G + Z_0} \tag{7}$$

where Z_{out} is the output impedance. And the reflection coefficients in (7) also depend on the scattering matrix parameters as (8).

$$\Gamma_{out} = S_{22} + \frac{S_{12}S_{21}\Gamma_G}{1 - S_{11}\Gamma_G} \tag{8}$$

The efficiency of the wireless power transfer can be deduced through the P_{in} (input power, coming into the two port network from the generator) and P_{out} (output power, going out from the two port network to the load). For the system in figure 3 from Ref.[1] the input and output power can be derived as

$$P_{in} = \frac{1}{2} \frac{|V_G|^2 R_{in}}{|Z_{in} + Z_G|^2}$$

$$P_L = \frac{1}{2} \frac{|V_G|^2 R_L |Z_{21}|^2}{|(Z_{11} + Z_G)(Z_{out} + Z_L)|^2}$$

(9)

where $R_{in}=Re\{Z_{in}\}$ and $R_L=Re\{Z_L\}$. In here "Re" stands for the real part of the complex number. From (9), a necessary condition for maximum power delivery from the generator to the connected system is given by (10).

$$Z_{in} = Z_G^*$$

(10)

Similarly the maximum output power could be delivered to the load when (11) holds.

$$Z_L = Z_{out}^*$$

(11)

Then in terms of S-parameters, the efficiency of wireless power transfer from Ref.[1] could be deduced as (12).

$$\eta_1 = \frac{\left(1 - |\Gamma_G|^2\right)|S_{21}|^2 \left(1 - |\Gamma_L|^2\right)}{\left|(1 - S_{11}\Gamma_G)(1 - S_{22}\Gamma_L) - S_{12}S_{21}\Gamma_G\Gamma_L\right|^2}$$

(12)

Here, if the load and generator impedances are matched to the reference impedance (i.e. $Z_G=Z_L=Z_0$), then from (7) and (8) reflection coefficients would be presented as (13).

$$\Gamma_L = \Gamma_G = 0 \text{ and } \Gamma_{in} = S_{11}, \Gamma_{out} = S_{22}$$

(13)

Substituting (13) to (12) the efficiency formula can be simplified as (14).

$$\eta_1 = |S_{21}|^2$$

(14)

2.2 Analysis based on the Coupled Mode Theory (CMT)
This section describes the CMT analysis. At first basic definitions for a simple LC circuit are introduced, and by sequentially adding losses, a full wireless power transfer system model including coupling effect is obtained.

2.2.1 Basic definitions
This book utilizes the resonance phenomena for the efficient wireless power transfer. Generally, resonance can take many forms: mechanical resonance, acoustic resonance, electromagnetic resonance, nuclear magnetic resonance, electron spin resonance, etc. The wireless power transfer relies on the electromagnetic resonance and it is discussed by means

Analysis of Wireless Power Transfer by Coupled Mode Theory (CMT) and Practical Considerations to Increase
Power Transfer Efficiency

35

of CMT. The presented coupled mode formalism is general and developed in more detail in
Ref. [2].

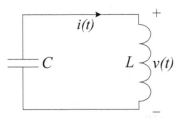

Fig. 4. An LC circuit.

Starting from the simple lossless ideal LC circuit (as a description of a single wireless power
transfer antenna or coil), presented in figure 4, the system description in a form of two
coupled first order differential equations (15) can be derived.

$$v = L\frac{di}{dt}$$
$$i = -C\frac{dv}{dt}$$
(15)

This system can be expressed by one second order differential equation

$$\frac{d^2v}{dt^2} + \omega^2 v = 0$$
(16)

where $\omega = \dfrac{1}{\sqrt{LC}}$ is a resonant frequency of LC circuit. Also, instead of a set of two coupled
differential equations in (15), two uncoupled differential equations can be derived as (18) by
defining the new complex variables defined in (17).

$$a_\pm = \sqrt{\frac{C}{2}}(v \pm j\sqrt{\frac{L}{C}}i)$$
(17)

By using a_+ and a_- (mode amplitude) it is possible to derive the system equations as (18).

$$\frac{da_+}{dt} = j\omega a_+$$
(18a)

$$\frac{da_-}{dt} = -j\omega a_-$$
(18b)

The square of mode amplitude is equal to the energy stored in a circuit. To verify this,
consider the following equations.

$$v(t) = |V|\cos(\omega_0 t)$$
(19)

$$i(t) = \sqrt{\frac{C}{L}}|V|\sin(\omega_0 t)$$
(20)

where $|V|$ is a peak amplitude of voltage in figure 4. Substituting (19) and (20) in (17), the mode amplitude, a, can be derived as

$$a_+ = \sqrt{\frac{C}{2}}(|V|\cos(\omega_0 t) + j|V|\sin(\omega_0 t)) = \sqrt{\frac{C}{2}}|V|e^{j\omega_0 t} \tag{21}$$

Hence,

$$|a_+|^2 = \frac{C}{2}|V|^2 = W \tag{22}$$

where W is the energy stored in the circuit. And similar procedure can be applied to a_-.
The main advantage of such transformations is the possibility to represent the system of coupled differential equations in a form of two uncoupled equations like (19). Moreover it is possible to use only one equation (18a) to describe the resonant mode, since the second one, (18b) is the complex conjugate form of (18a). Therefore in further analysis subscript + will be dropped and only equation (18a) will be used.

2.2.2 Lossy circuit
In above section a lossless circuit is considered and the equation for the mode amplitude is derived. But if the circuit is lossy, every practical circuit has loss, the equations must be modified. Such an electric circuit is presented in figure 5, where loss is presented by a resistance R.

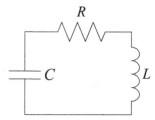

Fig. 5. Lossy LC circuit.

If the loss is small (as is generally true for the system of interest), then utilizing assumptions of perturbation theory equation (18a) could be expressed in a form of (23).

$$\frac{da}{dt} = j\omega a - \Gamma a \tag{23}$$

where Γ is the decay rate due to system losses. This decay rate can be calculated either from circuit (figure 5) analysis, or by applying the relation between power and the mode amplitude, a. Then, from (22), the decay rate can be used to explain the loss as (24).

$$\frac{d|a|^2}{dt} = \frac{dW}{dt} = -2\Gamma W = -P_{loss} \tag{24}$$

where P_{loss} is a power dissipated on the resistance, and it can be calculated by sequent equation.

$$P_{loss} = \frac{1}{2}|I|^2 R = \frac{WR}{L} \tag{25}$$

Thus unloaded (with no external load) quality factor Q of such a system can be expressed as (26).

$$Q = \omega \frac{W}{P_{loss}} = \frac{\omega}{2\Gamma} = \frac{\omega L}{R} \tag{26}$$

Hence decay rate Γ could be derived from (26) as (27).

$$\Gamma = \frac{R}{2L} \tag{27}$$

Other perturbations (coupling with other resonator, connection to the transmission line, etc) as discussed in Ref.[2] could be added in a similar manner to the intrinsic circuit loss.

2.2.3 Lossy circuit in the presence of a power source
Following the discussion in Ref.[2], when a power source exists (refer figure 6), (23) must be modified considering two factors:
1. Decay rate modification,
2. Mode amplitude excitation due to incident wave.
Decay rate is modified due to loss occurring not only in the coil alone, but also in a "waveguide" connecting source and antenna. Considering this, equation (23) can be modified in the following

$$\frac{da}{dt} = j\omega a - (\Gamma_{ext} + \Gamma)a \tag{28}$$

where Γ_{ext} represents the decay rate due to power escaping in a waveguide. Following the procedure presented in the previous section it can be seen that the product of external decay rate and energy stored in the scheme is linearly proportional to the power dissipated in a waveguide, so Γ_{ext} could be found through equation (15) as (29).

$$Q_{ext} = \frac{\omega}{2\Gamma_{ext}} \tag{29}$$

where Q_{ext} is an external system quality factor.

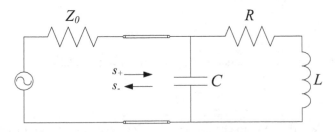

Fig. 6. Resonant circuit with external excitation and waveguide.

Further, considering the excitation by incident wave, (28) must be modified as (30).

$$\frac{da}{dt} = j\omega a - \left(\Gamma_{ext} + \Gamma\right)a + Ks_+ \tag{30}$$

where K expresses the degree of coupling between source and coil, and s_+ represents the incident wave incoming to the antenna. It is important to mention that $|s_+|^2$ has the meaning of input power, rather than energy as in case of $|a|^2$. Using reversibility property of Maxwell's equations, it is possible to show that

$$K = \sqrt{2\Gamma_{ext}} \tag{31}$$

Therefore, by applying (31) to (30), that the resonator mode is described by three parameters: self-resonant frequency ω, internal (Γ) and external (Γ_{ext}) decay rates as (32).

$$\frac{da}{dt} = j\omega a - \left(\Gamma_{ext} + \Gamma\right)a + \sqrt{2\Gamma_{ext}}\,s_+ \tag{32}$$

2.2.4 Coupling of two lossless resonant circuits

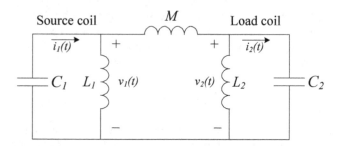

Fig. 7. Coupled resonators.

Presented formalism can be readily expanded to describe the coupling of two resonant circuits.

Suppose that a_1 and a_2 are the mode amplitudes of uncoupled lossless resonators with resonant frequencies ω_1 and ω_2, respectively. Then, if the resonators are coupled through some perturbation, (in the case of wireless power transfer it is the mutual inductance M), for the first resonator (18a) can be expressed as (33a).

$$\frac{da_1}{dt} = j\omega a_1 + k_{12}a_2 \tag{33a}$$

Consequently for the second resonator (18a) is transformed to (33b)

$$\frac{da_2}{dt} = j\omega a_2 + k_{21}a_1 \tag{33b}$$

where k_{12} and k_{21} are the coupling coefficients between the modes. Energy conservation law described in (34) provides a necessary condition for k_{12} and k_{21}.

Analysis of Wireless Power Transfer by Coupled Mode Theory (CMT) and Practical Considerations to Increase
Power Transfer Efficiency

39

$$\frac{d}{dt}(|a_1|^2 + |a_2|^2) = a_1 \frac{da_1^*}{dt} + a_1^* \frac{da_1}{dt} + a_2 \frac{da_2^*}{dt} + a_2^* \frac{da_2}{dt} =$$
$$= a_1^* k_{12} a_2 + a_1 k_{12}^* a_2^* + a_2^* k_{21} a_1 + a_2 k_{21}^* a_1^* = 0 \tag{34}$$

From (34), since "amplitudes of a_1 and a_2 can be set arbitrarily" [1], the coupling coefficients
must satisfy the sequent condition.

$$k_{12} + k_{21}^* = 0 \tag{35}$$

Similar to necessary conditions, exact values of k_{12} and k_{21} could be obtained through energy
conservation considerations. From the equation (33b) the power transferred from the first
resonator to the second resonator through the mutual inductance M (see figure 7) can be
evaluated as (36).

$$P_{21} = \frac{d|a_2|^2}{dt} = k_{21} a_1 a_2^* + k_{21}^* a_1^* a_2 \tag{36}$$

But from the electric circuit analysis for the circuit in figure 7 power flowing through M can
be expressed as (37).

$$P_{21} = i_2 M \frac{d(i_1 - i_2)}{dt} \tag{37}$$

Then, by introducing complex current envelope quantities $I_1(t)$, $I_2(t)$, an expression for
current $i_1(t)$ can be written as in the following.

$$i_1(t) = \frac{1}{2}(I_1(t)e^{j\omega_1 t} + I_1^*(t)e^{-j\omega_1 t}) \tag{38}$$

A similar procedure can be applied to the expression for current $i_2(t)$, then by substituting
both of these currents into (37), the expression for transferred power can be rewritten as

$$P_{21} = \frac{1}{4}(I_2(t)e^{j\omega_2 t} + I_2^*(t)e^{-j\omega_2 t})\left(\frac{d}{dt}\left(I_1(t)e^{j\omega_1 t} + I_1^*(t)e^{-j\omega_1 t} - I_2(t)e^{j\omega_2 t} - I_2^*(t)e^{-j\omega_2 t}\right)\right) \tag{39}$$

In (39), $\left(\frac{d}{dt}I_1\right)e^{j\omega_1 t}$ terms are much smaller than the $j\omega I_1$, so they can be ignored. By such a
approximation, (39) can be modified (40)

$$P_{21} = \frac{1}{4}\left(j\omega_1 M I_1 I_2^* e^{j(\omega_1 - \omega_2)t} - j\omega_1 M I_1^* I_2 e^{-j(\omega_1 - \omega_2)t}\right) \tag{40}$$

Then, comparing (40) with (36) and defining $a_n = \sqrt{\frac{L_n}{2}} I_n e^{j\omega_n t}$, where $n=1,2$, the coupling
coefficient can be derived as (41).

$$k_{21} = \frac{j\omega_1 M}{2\sqrt{L_1 L_2}} \tag{41}$$

Equation (41) yields that k_{21} is a purely complex number, so due to (35), k_{21} can be derived as (42).

$$k_{21} = k_{12} = jk = \frac{j\omega M}{2\sqrt{L_1 L_2}}$$
(42)

where ω must be interpreted as arithmetic mean $\frac{\omega_1 + \omega_2}{2}$ or geometric mean $\sqrt{\omega_1 \omega_2}$ of the corresponding coil's self resonance frequencies.

2.2.5 Full wireless power transfer system model

In figure 8 the scheme of the wireless power transfer system is depicted. Here Z_G stands for the internal impedance of the power source and d represents the distance between the source and load coils.

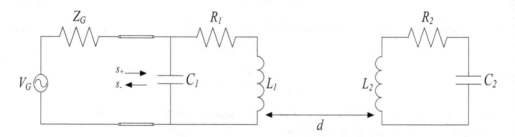

Fig. 8. Wireless power transfer scheme.

Incorporating the results presented in the above sections, the full model of such a system by CMT can be represented as follows,

$$\frac{da_1(t)}{dt} = j\omega_1 a_1(t) - \left(\Gamma_{ext_1} + \Gamma_1\right)a_1(t) + jka_2(t) + \sqrt{2\Gamma_{ext_1}}\,s_+(t)$$

$$\frac{da_2(t)}{dt} = j\omega_2 a_2(t) - \Gamma_2 a_2(t) + jka_1(t)$$
(43)

However, (43) still cannot express a full wireless power transfer system, because it does not contain term representing load. If the load is considered by defining $|s_{+n}(t)|^2$ as power ingoing to the object n ($n=1,2$) and $|s_{-n}(t)|^2$ as power outgoing from the resonant object n, the full system can be described as (44).

$$\frac{da_1(t)}{dt} = j\omega_1 a_1(t) - \left(\Gamma_{ext_1} + \Gamma_1\right)a_1(t) + jka_2(t) + \sqrt{2\Gamma_{ext_1}}\,s_{+1}(t)$$

$$\frac{da_2(t)}{dt} = j\omega_2 a_2(t) - \left(\Gamma_{ext_2} + \Gamma_2\right)a_2(t) + jka_1(t)$$

$$s_{-1}(t) = \sqrt{2\Gamma_{ext_1}}\,a_1(t) - s_{+1}(t)$$

$$s_{-2}(t) = \sqrt{2\Gamma_{ext_2}}\,a_2(t)$$
(44)

Analysis of Wireless Power Transfer by Coupled Mode Theory (CMT) and Practical Considerations to Increase
Power Transfer Efficiency

41

2.3 Finite-amount power transfer

Supposing "no source" and "no load" conditions, the system description (44) modifies to (45).

$$\frac{da_1(t)}{dt} = j\omega a_1(t) - \Gamma_1 a_1(t) + jka_2(t)$$

$$\frac{da_2(t)}{dt} = j\omega a_2(t) - \Gamma_2 a_2(t) + jka_1(t)$$

(45)

where index 1 stands for the source coil and 2 - for the load coil. (45) can be rewritten in another form as (46).

$$\dot{\vec{a}}(t) = A\vec{a}(t)$$

(46)

where $\vec{a}(t)$ is a vector which involves the mode amplitudes of the source coil and the load coil. And matrix A is defined as (47).

$$A = \begin{pmatrix} j\omega_1 - \Gamma_1 & jk \\ jk & j\omega_2 - \Gamma_2 \end{pmatrix}$$

(47)

Eigen-values of such system could be obtained through solving the characteristic equation, $det(A-sI)=0$. They can be presented in (48).

$$x_1 = j\frac{\omega_1 + \omega_2}{2} - \frac{\Gamma_1 + \Gamma_2}{2} + j\sqrt{\left(\frac{\omega_1 - \omega_2}{2} - j\frac{\Gamma_1 - \Gamma_2}{2}\right)^2 + k^2}$$

$$x_2 = j\frac{\omega_1 + \omega_2}{2} - \frac{\Gamma_1 + \Gamma_2}{2} - j\sqrt{\left(\frac{\omega_1 - \omega_2}{2} - j\frac{\Gamma_1 - \Gamma_2}{2}\right)^2 + k^2}$$

(48)

The eigen-values are complex and distinct so the solution for resonance modes will be in form of (49).

$$\vec{a}(t) = \begin{pmatrix} a_1(t) \\ a_2(t) \end{pmatrix} = c_1 V_1 e^{x_1 t} + c_2 V_2 e^{x_2 t}$$

(49)

where c_1 and c_2 are constants determined by the initial conditions and V_1, V_2 are eigen-vectors. For the simplicity of further discussion, the constant Ω_0 is defined as (50).

$$\Omega_0 = \sqrt{\left(\frac{\omega_1 - \omega_2}{2} - j\frac{\Gamma_1 - \Gamma_2}{2}\right)^2 + k^2}$$

(50)

By such a transformation, eigen-vectors can be calculated as follows,

$$V_1 = \begin{pmatrix} -\frac{1}{k}\left(\frac{\omega_2 - \omega_1}{2} + j\frac{\Gamma_2 - \Gamma_1}{2} - \Omega_0\right) \\ 1 \end{pmatrix}$$

$$V_2 = \begin{pmatrix} -\frac{1}{k}\left(\frac{\omega_2 - \omega_1}{2} + j\frac{\Gamma_2 - \Gamma_1}{2} + \Omega_0\right) \\ 1 \end{pmatrix}$$

(51)

Assuming that the source coil at time $t=0$ has energy $|a_1(0)|^2$ and at the same time energy contained in the load coil is $|a_2(0)|^2$. Then, by inserting (48),(50),(51) to (49) the resonant modes can be presented as (52).

$$a_1(t) = \left(a_1(0)\left(\cos(\Omega_0 t) + \frac{\Gamma_2 - \Gamma_1}{2\Omega_0}\sin(\Omega_0 t) - j\frac{\omega_2 - \omega_1}{2\Omega_0}\sin(\Omega_0 t) \right) + a_2(0)\frac{jk}{\Omega_0}\sin(\Omega_0 t) \right) \cdot e^{j\frac{\omega_1 + \omega_2}{2}t}e^{-\frac{\Gamma_1 + \Gamma_2}{2}t}$$

$$a_2(t) = \left(a_1(0)\frac{jk}{\Omega_0}\sin(\Omega_0 t) + a_2(0)\left(\cos(\Omega_0 t) - \frac{\Gamma_2 - \Gamma_1}{2\Omega_0}\sin(\Omega_0 t) + j\frac{\omega_2 - \omega_1}{2\Omega_0}\sin(\Omega_0 t) \right) \right) \cdot e^{j\frac{\omega_1 + \omega_2}{2}t}e^{-\frac{\Gamma_1 + \Gamma_2}{2}t}$$

(52)

Note that, for a special case when $\omega_1=\omega_2=\omega_0$ and $\Gamma_1=\Gamma_2=\Gamma_0$ (for the equal source and load antennas) and $|a_2(0)|^2=0$ (52) can be simplified to a set of equations as (53).

$$a_1(t) = a_1(0)\cos(kt)e^{j\omega_0 t}e^{-\Gamma_0 t}$$
$$a_2(t) = ja_1(0)\sin(kt)e^{j\omega_0 t}e^{-\Gamma_0 t}$$

(53)

Energy flow over time in a wireless power system (assuming zero initial load coil energy) described by (53) is illustrated in figure 9.

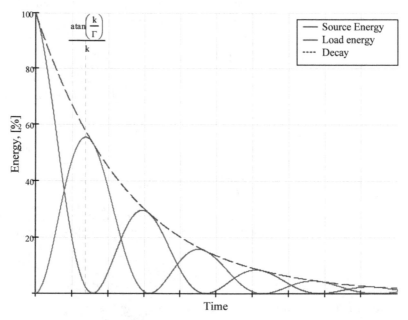

Fig. 9. Energy flow between the source and load coils over time.

Following the procedure described in Ref.[3], energy-transfer efficiency for a finite-amount of power transfer is determined by (54).

$$\eta = \frac{|a_2(t)|^2}{|a_1(0)|^2}$$

(54)

Then by defining $U = \dfrac{k}{\Gamma_0}$ and $T = \Gamma_0 t$ the time t^o which maximizes efficiency is calculated

through maximization of $\sin(UT) \cdot e^{-T} = \dfrac{|a_2(t)|}{|a_1(0)|}$. Utilizing Lagrange method, optimal time in

the sense of maximum energy transfer time can be calculated by the following equation

$$ T^o = \frac{\operatorname{atan}(U)}{U} \text{ or } t^o = \frac{\operatorname{atan}\left(\dfrac{k}{\Gamma}\right)}{k} \tag{55} $$

Then, substitution of (55) to (54) results in optimal transfer efficiency as (56).

$$ \eta\left(t^o\right) = \frac{U^2}{1+U^2} e^{\frac{-2\operatorname{atan}(U)}{U}} \tag{56} $$

It is important to note that (56) represents the optimal transfer efficiency which is only
dependent on U ("coupling to loss ratio" [3]) and tends to be unity when $U \gg 1$.
Although a finite-amount power transfer scheme is pure theoretical, one could consider it to
be a conceptual power transfer system, where the power source is connected to the source
coil at time $t=0$ for a very short time and then at time $t=t^o$ the load would drain energy from
the load coil. This procedure would repeat.

2.4 Wireless power transfer in the presence of source and load

This section considers the case when a power source is continuously connected to the source
coil through Γ_{ext_1} and the load is correspondingly connected to the load coil through Γ_{ext_2}.
Equations for wireless power transfer defining such a case was already described in 2.2.5.
Here, they are repeated for the convenience of future system efficiency calculations.

$$ \frac{da_1(t)}{dt} = j\omega_1 a_1(t) - \left(\Gamma_{ext_1} + \Gamma_1\right)a_1(t) + jka_2(t) + \sqrt{2\Gamma_{ext_1}}\, s_{+1}(t) $$

$$ \frac{da_2(t)}{dt} = j\omega_2 a_2(t) - \left(\Gamma_{ext_2} + \Gamma_2\right)a_2(t) + jka_1(t) \tag{44} $$

$$ s_{-1}(t) = \sqrt{2\Gamma_{ext_1}}\, a_1(t) - s_{+1}(t) $$

$$ s_{-2}(t) = \sqrt{2\Gamma_{ext_2}}\, a_2(t) $$

Similar to the discussion in Ref.[3], it assumes that the excitation frequency is fixed and
equal to ω. And, the field amplitudes has the form of (57).

$$ s_{+1}(t) = S_{+1} e^{j\omega t} \tag{57} $$

where S_{+1} is some constant determined by the amplitude of incident wave. Then it is
possible to express the system (44) in matrix form as follows,

$$ \dot{\vec{a}}(t) = \begin{pmatrix} j\omega_1 - \Gamma_1 & jk \\ jk & j\omega_2 - \Gamma_2 \end{pmatrix} \vec{a}(t) + \begin{pmatrix} \sqrt{2\Gamma_{ext_1}} \\ 0 \end{pmatrix} s_{+1}(t) \tag{58} $$

$$ \vec{y}(t) = (1\ 1)\vec{a}(t) $$

The transfer function of the system (58) can be derived as (59).

$$\hat{g}(s) = \frac{\sqrt{2\Gamma_{ext_1}}\,(s - j\omega_2 + \Gamma_2 - jk)}{s^2 - s(j\omega_1 + j\omega_2 + \Gamma_1 + \Gamma_2) - \omega_1\omega_2 - j\omega_1\Gamma_2 - j\omega_2\Gamma_1 + \Gamma_1\Gamma_2 + k^2} \tag{59}$$

And it has a pair of poles $s_{1,2}$ defined in (60).

$$s_{1,2} = -\frac{\Gamma_1 + \Gamma_2}{2} + j\frac{\omega_1 + \omega_2}{2} \pm \frac{1}{2}\sqrt{\left(j(\omega_1 - \omega_2) + \Gamma_2 - \Gamma_1\right)^2 - 4k^2} \tag{60}$$

From (60), it is apparent that for the typical parameters of interest, poles of transfer function (59) always have negative real parts, so the system (58) is Boundary Input Boundary Output (BIBO) stable. BIBO stability together with (57) imply that the response of a linear system ($\vec{a}(t)$) will be at the same frequency ω as the input and could be represented in the form of (61).

$$a_{1,2}(t) = A_{1,2}e^{j\omega t} \tag{61}$$

A similar discussion for $s_{-1,2}(t)$ leads to the conclusion that the expression for the outgoing waves has the same form as (61) and can be represented as (62).

$$s_{-1,2}(t) = S_{-1,2}e^{j\omega t} \tag{62}$$

By substituting (61) and (62) into (44) and taking into account the time derivative of (62) such as $\dot{a}_{1,2}(t) = j\omega A_{1,2}e^{-j\omega t} = j\omega a_{1,2}(t)$, it is possible to calculate transmission coefficient S_{21} (from the scattering matrix). By defining $\delta_{1,2} = \omega - \omega_{1,2}$, $D_{1,2} = \dfrac{\delta_{1,2}}{\Gamma_{1,2}}$, $U_{1,2} = \dfrac{\Gamma_{ext_{1,2}}}{\Gamma_{1,2}}$ and $U = \dfrac{k}{\sqrt{\Gamma_1\Gamma_2}}$, S_{21} can be expressed as (63).

$$S_{21} = \frac{S_{-2}}{S_{+1}} = \frac{2jk\sqrt{\Gamma_{ext_1}\Gamma_{ext_2}}}{(\Gamma_1 + \Gamma_{ext_1} + j\delta_1)(\Gamma_2 + \Gamma_{ext_2} + j\delta_2) + k^2} = \frac{2jU\sqrt{U_1 U_2}}{(1 + U_1 + jD_1)(1 + U_2 + jD_2) + U^2} \tag{63}$$

Similarly, the reflection coefficient S_{11}, can be determined ("field amplitude reflected to the generator" [3]).

$$S_{11} = \frac{S_{-1}}{S_{+1}} = \frac{(\Gamma_1 - \Gamma_{ext_1} + j\delta_1)(\Gamma_2 + \Gamma_{ext_2} + j\delta_2) + k^2}{(\Gamma_1 + \Gamma_{ext_1} + j\delta_1)(\Gamma_2 + \Gamma_{ext_2} + j\delta_2) + k^2} = \frac{(1 - U_1 + jD_1)(1 + U_2 + jD_2) + U^2}{(1 + U_1 + jD_1)(1 + U_2 + jD_2) + U^2} \tag{64}$$

Other parameters of the scattering matrix can be obtained simply by interchanging notations 1 and 2 due to the reciprocity property.

The purpose of this present discussion is to find the optimal parameters which will maximize the efficiency as determined by the maximum ratio of power between the power "obtained" by the load and the power "sent" by the source. The conditions can be found solving the following problems:

Analysis of Wireless Power Transfer by Coupled Mode Theory (CMT) and Practical Considerations to Increase
Power Transfer Efficiency

45

1. Power transmission efficiency maximization (i.e. $|S_{21}|^2 \rightarrow \max$)

2. Power reflection minimization at generator side/load side (i.e. $|S_{11}|^2 \rightarrow \min$)

2.4.1 Maximization of power transmission efficiency

As discussed in the preceding section, to obtain maximum transmission efficiency, $|S_{21}|^2$ must be maximized. From (63), it can be determined that this expression is symmetric by interchanging 1↔2. Therefore, optimal values for D_1 and D_2 must be the same as like U_1 and U_2. By defining $D_0=D_1=D_2$ and $U_0=U_1=U_2$, (63) is transformed into (65).

$$S_{21} = \frac{2jUU_0}{(1+U_0+jD_0)^2 + U^2} \qquad (65)$$

From here, by utilizing Lagrange method it is possible to obtain optimal values for D_0 (for fixed values of U and U_0). Power transmission efficiency could be maximized by setting D_0 as (66).

$$D_0^* = \begin{cases} \pm\sqrt{U^2 - (1+U_0)^2} & if \ U > 1+U_0 \\ 0, & if \ U \le 1+U_0 \end{cases} \qquad (66)$$

Further, with this optimal D_0^*, by the similar procedure it is possible to show that the optimal value for U_0 is expressed as (67).

$$U_0^* = \sqrt{1+U^2} \qquad (67)$$

From figure 10 cited from Ref.[3], it can be seen that the matching of condition in (67) is very important to maximize the power transfer efficiency. In this figure the system efficiency

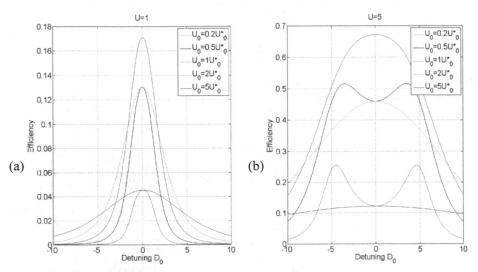

Fig. 10. Efficiency of wireless power transfer scheme as a function of U_0 and detuning D_0[3].

versus frequency for different values of $\dfrac{U_0}{U_0^*}$ is presented. From this graph maximum efficiency is obtained when $U_0 = U_0^*$, and the physical meaning of "over-coupled" defined as the system where $U_0 > U_0^*$ and "under-coupled" as $U_0 < U_0^*$ becomes clear.

From (66), the optimal value of D_0 can only be 0. So substituting this result together with (67) into (65), the optimal power transmission efficiency is given by (68).

$$\eta^* = \eta\left(D_0^*, U_0^*\right) = \left(\frac{U}{1+\sqrt{1+U^2}}\right)^2 \tag{68}$$

From (68), it is important to mention that efficiency is again just a function of U, where $U = \dfrac{k}{\sqrt{\Gamma_1 \Gamma_2}}$, and it approaches unity as U increases to infinity.

Note that, in a real system, the strict satisfaction of all the optimal conditions is very difficult or even impossible. For example, from (67) it can be said that the optimal U_0 should be set by U. However U_0 is determined by the source and load and is therefore not controllable. Moreover, U is dependent on the distance of power transfer. Referring to figure 10(b), when the system is over-coupled to achieve maximum efficiency (at two peaks on the figure) the frequency of the power source must not be the same with the resonant frequencies of objects ($D_0^* \neq 0$). That is why it is significant to find these new optimal power supply frequencies. The optimal frequency which will maximize the power transmission efficiency in this case is derived as (69).

$$\omega_\pm = \begin{cases} \dfrac{\omega_1 \Gamma_2 + \omega_2 \Gamma_1}{\Gamma_1 + \Gamma_2} \pm \dfrac{2\sqrt{\Gamma_1 \Gamma_2}}{\Gamma_1 + \Gamma_2}\sqrt{k^2 - (\Gamma_1 + \Gamma_{ext_1})(\Gamma_2 + \Gamma_{ext_2})}, & \text{if } \dfrac{k}{\Gamma_1 \Gamma_2} > \sqrt{\left(1 + \dfrac{\Gamma_{ext_1}}{\Gamma_1}\right)\left(1 + \dfrac{\Gamma_{ext_2}}{\Gamma_2}\right)} \\[2em] \dfrac{\omega_1 \Gamma_2 + \omega_2 \Gamma_1}{\Gamma_1 + \Gamma_2}, & \text{if } \dfrac{k}{\Gamma_1 \Gamma_2} \leq \sqrt{\left(1 + \dfrac{\Gamma_{ext_1}}{\Gamma_1}\right)\left(1 + \dfrac{\Gamma_{ext_2}}{\Gamma_2}\right)} \end{cases} \tag{69}$$

In (69), the first equation for the power supply frequency ω corresponds to the over-coupled system, the second – to under-coupled system. To indicate this regime of operation given by (69) the term "semi-optimal conditions" can be used. Note, that if we define $\delta_{sp} = \omega_+ - \omega_-$ (splitting of frequencies), then for the same resonant objects (i.e. $\Gamma_1 = \Gamma_2 = \Gamma_0$ and $\Gamma_{ext_1} = \Gamma_{ext_2} = \Gamma_{ext_0}$) splitting of frequencies will be described as (70).

$$\delta_{sp} = 2\sqrt{k^2 - \left(\Gamma_0 + \Gamma_{ext_0}\right)^2} \tag{70}$$

And this becomes a criterion for the coupling coefficient k value calculation.

2.4.2 Minimization of power reflection

Minimization of the reflection of the powers can be achieved by minimization of $|S_{11}|^2$ [3]. From (64), conditions of $|S_{11}|^2$ minimization are formulated as (71).

$$D_{1,2}^*, U_{1,2}^* : S_{11,22} = 0 \Rightarrow (1 \mp U_1 + jD_1)(1 \pm U_2 + jD_2) + U^2 = 0 \tag{71}$$

Similar to the discussion regarding equation (71) in 2.4.1, it is symmetric by interchanging $1 \leftrightarrow 2$. And it can be said that the optimal values for $D_{1,2}$ are the same and equal to some new[1] D_0^* (as with $U_{1,2}$). Substituting these considerations into (71) results in (72).

$$(1+ jD_0)^2 - U_0^2 + U^2 = 0 \tag{72}$$

Equation (72) (which equally represents the impedance matching problem in terms of CMT) results in the optimal conditions for D_0^* and U_0^*, as (73).

$$D_0^* = 0$$
$$U_0^* = \sqrt{1+U^2} \tag{73}$$

Accordingly, both power transmission efficiency maximization and power reflection minimization at both the generator side and load side lead to the same set of conditions described by (73). Therefore, the following does not divide the term "efficiency" into parts, instead it includes both meanings.

2.5 Efficient wireless power transfer: Scattering matrix analysis vs. CMT results
In above sections, the conditions for efficiency maximization, based on two different approaches (scattering matrix analysis and CMT), are presented. From section 2.1, the scattering matrix analysis, the three conditions used to maximize efficiency are given by (74):

$$\left. \begin{array}{l} Z_{in} = Z_s^* \\ Z_{out} = Z_L^* \end{array} \right\} \text{"No reflection" condition} \left. \begin{array}{l} \\ \\ \\ \end{array} \right\} Z_{in} = Z_L^*. \tag{74}$$
$$Z_s = Z_L \text{ Symmetric condition}$$

From section 2.2, for the CMT, the two conditions which maximize efficiency are given by (75).

$$D_1^* = D_2^* = D_0^* = 0$$
$$U_1^* = U_2^* = U_0^* = \sqrt{1+U^2} \tag{75}$$

The variables that are used to describe the optimal conditions in (74) and (75) are different. So the question naturally arises "do these conditions contradict to each other?" To answer this question, an equivalent circuit of a wireless power system connected to the source and load can be considered as shown in figure 11.
In figure 11 R_{ac1} and R_{ac2} represent coil losses and M stands for the mutual inductance between the coils. In addition, R_s is an internal source impedance and R_L , the load impedance. C_{ext1} and C_{ext2} represent the external capacitors connected to the antennas (internal capacitance of the coil is neglected). Supposing that $R_s = R_L$ and the coils are the same (due to the equivalence of coils, subscripts 1,2 are omitted in following discussion).

---

[1] It would be shown that optimal conditions for sections 2.4.1 and 2.4.2 are absolutely same. So it is not required to introduce a new variable for current discussion.

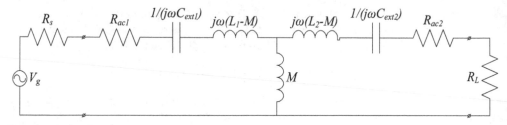

Fig. 11. Equivalent circuit of wireless power transfer in lumped parameters.

This assumption automatically leads to the satisfaction of the symmetric condition given by (74) and provides that $D_1 = D_2$ and $U_1 = U_2$.

With these considerations, condition (74) for the circuit in figure 11 can be developed as (76).

$$Z_{in} = R + j\omega((L-M)) + \frac{1}{j\omega C} + \frac{j\omega M\left(R + R_L + j\omega(L-M) + \frac{1}{j\omega C}\right)}{R + R_L + j\omega L + \frac{1}{j\omega C}} = R_L$$

or   (76)

$$R^2 - R_L^2 + \omega^2\left(M^2 - L^2\right) + 2j\omega LR + \frac{2R}{j\omega C} + \frac{2L}{C} = 0$$

And if the resonance type of operation is considered (i.e. $\omega = \frac{1}{\sqrt{LC}}$), (76) can be simplified to

$$R_L^2 = R^2 + \frac{M^2}{LC} \qquad (77)$$

Meanwhile, from (75), the assumption of resonant operation leads to $D_1^* = D_2^* = D_0^* = 0$. In order to represent the second condition of CMT in terms of figure 11 (78) can be derived.

$$U = \frac{R_L}{R} \text{ and } U = \frac{\omega M}{R} \qquad (78)$$

Substituting (78) into the second equation of (75) results in the following,

$$R_L^2 = R^2 + \frac{M^2}{LC} \qquad (79)$$

Equation (79) is exactly same to (77) which was obtained through scattering matrix analysis of the circuit.

As shown above, both theories give the same result so any of them can be used as a basis for maximum power transfer research. In here, CMT would be used to explain the phenomenon of the wireless power transfer, because S-parameters (scattering matrix's elements) are not easily predicted, and because the system design based on scattering matrix is problematic. So, the usage of S-parameters is restricted to the experimental verification of results, due to the simple *experimental* calculation of the S-matrix by a network analyzer. Also, some papers have proposed the wireless power system can be described by the conventional circuit analysis. However, this method is too complicated to utilize in this field (for example, in the case of indirect feeding). Moreover, such an analysis is not universal and will vary from a circuit to another circuit. For example, if external capacitance is added to the coil, the whole model must be recalculated. In contrast, CMT provides an accurate and convenient way of modeling of the system and, as it was presented in section 2.4, expresses the system as a set of linear differential equations given by (80).

$$\dot{\vec{a}}(t) = \begin{pmatrix} j\omega_1 - \Gamma_1 & jk \\ jk & j\omega_2 - \Gamma_2 \end{pmatrix} \vec{a}(t) + \begin{pmatrix} \sqrt{2\Gamma_{ext\,1}} \\ 0 \end{pmatrix} s_{+1}(t) \tag{80}$$

$$\vec{y}(t) = \begin{pmatrix} 1 & 1 \end{pmatrix} \vec{a}(t)$$

Here, it is important to note that CMT is based on several assumptions. First, the internal loss of coils is comparatively small (to satisfy the perturbation theory requirements). Second, the frequency range of operation is narrow enough to assume inductance and internal capacitance of the coils be constant. Third, "overall field profile can be described as a superposition of the modes due to each object" [4].

3. Analysis of a real system by CMT & practical considerations to increase efficiency

As presented in a section 2, $U = \dfrac{k}{\sqrt{\Gamma_1 \Gamma_2}}$, (69) is the figure of merit to maximize the efficiency of the power transfer. Also, in the same section, the quality factor of each coil is defined as $Q = \dfrac{\omega}{2\Gamma}$, (26). By rearranging two equations, the following relationship can be obtained.

$$U = \frac{k}{\sqrt{\Gamma_1 \Gamma_2}} = K_U \sqrt{Q_1 Q_2} \tag{81}$$

where K_U is a coupling factor between two coils ($K_U = \dfrac{M}{\sqrt{L_1 L_2}}$) and Q_1 and Q_2 are the quality factors of the source and load coils, respectively. From equation (81), to achieve a high figure of merit U (i.e. to get maximum efficiency), one should use resonant coils with a high quality factor (or with a low intrinsic loss rate Γ). Another way to increase efficiency is to make a tight coupling between two resonators. However, in practical applications there is always a geometrical limit of object size, and it applies restriction to K_U and Q_n magnification. In the following sections, ways of calculating Q_n and K_U will be presented together with ways to increase the figure of merit U.

3.1 Capacitive-loaded helical coils

In Ref.[3], [5] and [6], wireless power transfer by resonant coils without external capacitance was widely discussed and practical implementation was shown. For example, in Ref.[5] wireless power transfer between coils over 2 meters distance with a maximum 40% efficiency was reported. But the frequencies of such a power transfer exceed 10MHz, which may be costly in designing such a high frequency power source. Here, the goal is to reduce the operational frequency of power transfer. One solution to reduce the resonant frequency of object is to add a large external capacitor ("large external capacitor" means that its value exceeds the coil's self-capacitance). Such systems have been discussed in Ref.[3] and Ref.[7]. In figure 12, a capacitive-loaded helical coil is presented. In order to calculate the quality factor of such a coil the inductance, self capacitance, and resistance must be evaluated.

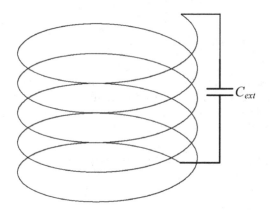

Fig. 12. Diagram of capacitive-loaded helical coil.

3.1.1 Inductance, self-capacitance and resistance of a helical coil

Following the procedure presented in Ref.[8], the calculation of the quality factor starts from the inductance evaluation of the helical coil. The concept of the lumped inductance is strictly applicable only at low frequencies, i.e. when the length of wire consisting coil is much smaller than the wavelength of the frequency. This means that waves enter into the input terminal of the antenna and come out from the output terminal with virtually no phase difference. Obviously, this is not true at high frequencies. When speaking about the inductance of a coil at high frequencies, it is useful to segregate the discussion into two parts: external inductance due to the magnetic energy stored in a surrounding medium and internal inductance due to the magnetic field stored inside the wire itself.

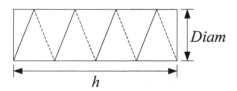

Fig. 13. Diagram of current-sheet inductor.

Inductance calculations are normally started from analysis of a so-called current-sheet inductor, as shown in figure 13. Such a solenoid is assumed to have an infinitely thin conducting wound wire (one layer) with no spacing between the turns (nevertheless the turns are electrically isolated). The main characteristic of such a coil is that at low frequencies they have a uniform magnetic field distribution along their length. If all these assumptions are satisfied, the expression for the inductance can be expressed as (82):

$$L_s = \frac{\mu \pi Diam^2 n^2}{4h} \tag{82}$$

where $Diam$ is the coil diameter, n – number of turns, h – height of the coil and μ is magnetic permeability, which in the absence of a core reduces to μ_0.

The current-sheet inductor is a theoretical model, but by using this model the formula for the real inductors could be obtained after the introduction a few modifications. The modifications can be divided into two parts: frequency independent and frequency dependent. The first consists of a field non-uniformity correction coefficient (k_L), a self induction correction coefficient for a round wire (k_s) and a mutual inductance correction coefficient for a wound wire (k_m). The frequency dependent parameters which must be taken in consideration are $Diam$ (effective loop diameter), internal inductance (L_i) and self-capacitance (C).

3.1.2 Frequency independent modifications

In figure 14 a real coil is depicted, where wires forming the coil have some finite size (round wire with radius a and height h) and there is non-zero spacing between the coil turns (p is the pitch of the coil)

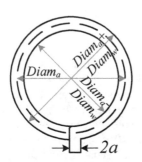 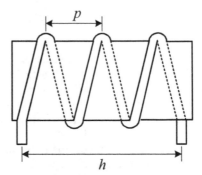

Fig. 14. Inductor with finite size, made from round wire [8].

When coil length is comparable to its diameter, the assumption of uniform field distribution is not valid. In 1909 H.Nagaoka introduced coefficient k_L [9] to consider this non-uniformity (83).

$$k_L = \frac{2h}{\pi Diam}\left|\frac{\left(\ln\left(\frac{4Diam}{h}\right)-\frac{1}{2}\right)\left(1+0.383901\left(\frac{h}{Diam}\right)^2+0.017108\left(\frac{h}{Diam}\right)^4\right)}{1+0.258952\left(\frac{h}{Diam}\right)^2}\right.$$
$$\left.+0.093842\left(\frac{h}{Diam}\right)^2+0.002029\left(\frac{h}{Diam}\right)^4-0.000801\left(\frac{h}{Diam}\right)^6\right| \tag{83}$$

With this correction, (82) can be modified to the following equation.

$$L_s = \frac{\mu\pi Diam^2 n^2}{4h}k_L \tag{84}$$

However, for the real coils, a coefficient, k_s, to consider the round shape of the wire must be added, and, moreover, another coefficient, k_m, to consider the mutual inductance between the adjacent turns must be included. To take these factors into account, E.B.Rosa [10] modified (84) to a new expression (85).

$$L = L_s - \frac{\mu n Diam}{2}(k_s + k_m) \tag{85}$$

where

$$k_s = \frac{3}{2}-\ln\left(\frac{p}{a}\right)$$
$$k_m = \ln(2\pi)-\frac{3}{2}-\frac{\ln(n)}{6n}-\frac{0.33084236}{n}-\frac{1}{120n^3}+\frac{1}{504n^5} \tag{86}$$
$$-\frac{0.0011923}{n^7}+\frac{0.0005068}{n^9}$$

and μ, in the absence of a core, can be considered as just μ_0.

3.1.3 Frequency dependent modifications

As discussed in a previous section, figure 14 represents a coil with finite dimensions. In conventional formulas, the diameter of a coil which is used to calculate inductance is equal to $Diam=Diam_a$, but this is not right. There are several reasons: even at low frequencies conduction pass outside the coil (i.e. $Diam_a+Diam_w$) is longer than the inside pass ($Diam_a-Diam_w$), also the stretching of wire, while producing the coil, increases the resistivity of the outside pass. So the effective loop diameter must be modified, due to the tendency of the current density to move towards the inside edge of the coil. This is particularly important due to $Diam$ in (82) where the inductance is proportional to the square of $Diam$. In another study in Ref.[8], the effective current diameter for a low frequency case was evaluated as (87).

$$Diam_0 = Diam_a\left(1-\left(\frac{2a}{Diam_a}\right)^2\right) \tag{87}$$

But at higher frequencies two more effects must be taken into account: skin effect and proximity effect. The physics which lies in the basis of these effects is quite complex and direct calculation of the effective current diameter is almost impossible, but a semi-empirical formula had been developed in Ref.[8] and the effective diameter, $Diam$, could be obtained from (88).

$$Diam_{\infty} = \frac{Diam_0 + \dfrac{2Diam_{min}}{\left(\dfrac{p}{2a}-1\right)}}{1+\dfrac{2}{\left(\dfrac{p}{2a}-1\right)}}, \text{ where } Diam_{min} = Diam_a - 2a + \frac{4a}{n}$$

$$Diam = \Theta\left(Diam_0 - Diam_{\infty}\right) + Diam_{\infty}, \text{ where } \Theta = 2\delta_i\left(1-\exp\{-\left[\frac{a}{2\delta_i}\right]^{3.8}\}\right)^{1/3.8}(1-y), \quad (88)$$

$$y = \frac{0.0239}{\left(1+1.67\left(z^{0.036}-z^{-0.72}\right)^2\right)^4}, \ z = \frac{a}{2.552\delta_i}, \ \delta_i = \sqrt{\frac{\rho}{\pi f \mu_0}}$$

In (88), ρ is the resistivity of the material, f, the frequency of operation, and δ_i the skin depth. This formula comes from the observation that $Diam_{min}$ is the absolute minimum effective diameter of the helical coil, so $Diam_{\infty}$ (effective diameter at very high frequency) must lie somewhere between $Diam_0$ and $Diam_{min}$. Further, Θ is used as the weighting factor to evaluate the effective diameter on the specified frequency.

Internal inductance is an imaginary counterpart of the skin effect [8], and it decreases rapidly as the frequency increases. It is also proportional to wire-length (and consequently to number of turns n). However, external inductance is proportional to $n,^2$ so the effect of internal inductance can be considered for short coils only. Again, from Ref.[8] internal inductance is approximated as (89).

$$L_i = \frac{\mu_0 \delta_i}{4\pi a}\left(1-\exp\{-\left[\frac{a}{2\delta_i}\right]^{3.8}\}\right)^{1/3.8}(1-y)l \qquad (89)$$

where l is the wire length, which is equal to $l = \sqrt{(\pi n Diam)^2 + h^2}$.

Finally, taking into consideration of all of the frequency-dependent and independent corrections, the self inductance of the coil presented in figure 14 can be calculated using (90).

$$L = L_s - \frac{\mu n Diam}{2}(k_s + k_m) + L_i \qquad (90)$$

As discussed at the beginning of 3.1.1, at high frequencies, direct modeling of a coil by means of lumped inductance is not acceptable. But it is possible to represent the coil by lossy (R_{ac}) inductance (L from (90)) and parallel capacitance (C). With respect to the antenna terminals, the equivalent electric circuit can be presented as figure 15.

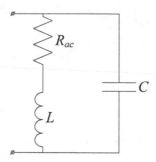

Fig. 15. Coil equivalent circuit.

However, R_{ac} and C are also frequency-dependent variables. Analytical expressions to define their values are presented in the following discussion.

In 1947 R.G. Medhurst [11], by numerous experiments found a semi-empirical formula which calculates the self capacitance and high frequency resistance of single layer solenoids. His formula, however, was acceptable only for coils with solid polyester cores. So, for general cases (any core material), his formula must be modified, resulting in (91). [11]

$$C = \frac{4\varepsilon_0\varepsilon_{rx}h}{\pi}\left(1+k_c\left(1+\frac{\varepsilon_{ri}}{2\varepsilon_{rx}}\right)\right)\left(1+\left(\frac{h}{\pi n Diam}\right)\right)^2, \text{ where}$$

$$k_c = 0.717439\left(\frac{Diam}{h}\right) + 0.933048\left(\frac{Diam}{h}\right)^{3/2} + 0.106\left(\frac{Diam}{h}\right)^2$$

(91)

In (91) ε_{rx} is the relative permittivity of the medium external to the solenoid and ε_{ri} that inside the solenoid, respectively.

Also in Ref.[11], the resistance of solenoid coils was widely discussed. It was shown that typically, four components form the resistance, i.e. the R_{dc} component, the component due to the skin effect (Θ), the component due to the proximity effect (Ψ), and for very high frequencies a radiated resistance component (R_r) must be added.

$$R_{ac} = R_{dc}\Theta\Psi + R_r, \text{ where}$$

$$R_r = \sqrt{\frac{\mu_0}{\varepsilon_0}}\left(\frac{\pi}{12}n^2\left(\frac{\omega Diam}{2c}\right)^4 + \frac{2}{3\pi^2}\left(\frac{\omega h}{c}\right)^2\right)$$

(92)

In (92), c is the speed of light, $R_{dc} = \frac{\rho l}{\pi a^2}$ and $\Theta = \frac{a^2}{\left(2a\delta_i - \delta_i^2\right)}$.

Both the R_{dc} and Θ components were studied and their formulas were presented in the previous paragraph. But the proximity effect factor is not easily derived. Through his experiments, Medhurst formed a table of coefficients Ψ, based on the geometric properties of coils (Table 1).

Here, it is important to note that table 1 is applicable only when the number of turns is large. For smaller numbers of turns, a weighting factor must be included, as was shown in Ref.[8], instead of (92) the high frequency resistance can be calculated through the following equation.

$$R_{ac} = R_{dc}\left(1 + \frac{(\Xi - 1)\Psi\left(n - 1 + \frac{1}{\Psi}\right)}{n}\right) + R_r \tag{93}$$

Note, however, that all of the formulas presented in section 3.1.1 are applicable only in case when the operational frequency is considerably smaller then the self-resonant frequency of the coil.

p/(2a) \ h/D	1	1.111	1.25	1.43	1.66	2	2.5	3.33	5	10
0	5.31	3.73	2.74	2.12	1.74	1.44	1.20	1.16	1.07	1.02
0.2	5.45	3.84	2.83	2.20	1.77	1.48	1.29	1.19	1.08	1.02
0.4	5.65	3.99	2.97	2.28	1.83	1.54	1.33	1.21	1.08	1.03
0.6	5.80	4.11	3.10	2.38	1.89	1.60	1.38	1.22	1.10	1.03
0.8	5.80	4.17	3.20	2.44	1.92	1.64	1.42	1.23	1.10	1.03
1	5.55	4.10	3.17	2.47	1.94	1.67	1.45	1.24	1.10	1.03
2	4.10	3.36	2.74	2.32	1.98	1.74	1.50	1.28	1.13	1.04
4	3.54	3.05	2.60	2.27	2.01	1.78	1.54	1.32	1.15	1.04
6	3.31	2.92	2.60	2.29	2.03	1.80	1.56	1.34	1.16	1.04
8	3.20	2.90	2.62	2.34	2.08	1.81	1.57	1.34	1.165	1.04
10	3.23	2.93	2.65	2.27	2.10	1.83	1.58	1.35	1.17	1.04
∞	3.41	3.11	2.815	2.51	2.22	1.93	1.65	1.395	1.19	1.05

Table 1. Proximity factor Ψ .

3.1.4 Maximization of helical coil quality factor

From (81), to maximize the power transfer efficiency, the quality factor of each coil must be maximized. Following (26), the quality factor is inversely proportional to the antenna resistance. As given by (93), resistance consists of four parts. The radiated resistance for typical parameters of interest is very small and could be neglected in the optimization process. The DC part can be suppressed through a series of steps: conductivity maximization, decreasing wire length and cross-sectional area magnification (well known in the case of DC).

However, wireless power transfer requires several *MHz* frequencies of operation. At such frequencies, the proximity effect and the skin effect become extremely important. For example, considering Table 1, when there is no spacing between the turns, coil resistance increases approximately 5 times. Fortunately, it could be easily suppressed by increasing spacing between the coils. Also from Table 1, when pole pitch exceeds wire diameters by more than 10 times, the proximity effect is practically negligible.

However, skin effect is not easily reduced. Normally, for example, in high-frequency transformers, Litz wire is used to minimize the skin effect. If only skin effect component is considered in (2.93), then the AC resistance can be expressed as (94)[8].

$$R_{ac} = \frac{\rho}{\delta_i}\left(\frac{l}{2\pi a}\right) \propto \frac{1}{P} \tag{94}$$

where P is the wire perimeter. So now, in contrary to the DC case, resistance is inversely proportional to the perimeter, not the cross-sectional area of a wire. This immediately provides that a wire with a circular cross-section is the worst case for resistance minimization. However, circular wires are the most common in coil manufacturing. So, as it was discussed in order to reduce skin effect, one could use Litz wire. Litz wire is made by twisting several insulated wires in such a way that each conductor has equal possibility to appear on the edge of a bundle. The idealized bundle's cross-section of seven wires is shown in figure 16.

$2a$

Fig. 16. Ideal cross-section of the Litz wire.

The wire presented in the figure 16 has a diameter equal to $2a$, and according to the picture, each conductor forming this Litz-wire has a diameter of $\frac{2a}{3}$. This means that the total perimeter of Litz-wire (as a sum of the perimeters of each conductor) is equal to $\frac{14\pi a}{3}$, but the perimeter of a normal wire with the same cross-sectional area would be $2\pi a$. This means that the resistance of the Litz-wire would be around 2.33 times smaller, according to (94). So Litz wire seems to be the perfect candidate to increase the total efficiency of a wireless power transfer system [3]. But it is not the case. The reason lies in the non-considering proximity effect in the adjacent conductors of the Litz wire. The proximity effect results in current redistribution, causing it to flow only on the outward-facing surfaces of the conductors forming the litz wire. With increasing frequency, this effect will be intensified.

Number of turns, n	4	4	3.7	4	4	4
Coil height, h [cm]	0.65	12	0.65	0.68	12	11
Wire type	Magnetic	Magnetic	Magnetic	Litz	Litz	Litz
Coil resonant frequency, f_1 [MHz]	4.24	5.28	4.453	4.46	5.38	5.46
Cross sectional area, S [mm²]	2.08	2.08	2.08	2.26	2.26	3.22
Resistance, R [Ohm]	2.21	1.2	1.99	4.6	2.37	2.57
Quality factor, Q	156.3	204	181.8	74.07	116.28	90.9

Table 2. Influence of Litz-wire and proximity effect on Q of single coil (C_{ext}=100pF).

For a practical verification of the proximity effect and Litz wire influences on coil resistance a set of experiments had been conducted. The results are tabulated in table 2. From the table two important conclusions can be made. First (blue box), due to the proximity effect resistance is increased by around two times. This experiment proves the possibility to reduce the proximity effect by simply increasing the spacing between the turns. Second (red box), Litz wire shows less performance on higher frequencies than magnetic wire with circular cross-section (note that frequencies of operation are nearly the same). In literature Litz-wire area of application is claimed to be about from *50KHz to 1-3MHz*. Our experiments show that Litz-wire shows worse performance starting from ~*2MHz*.

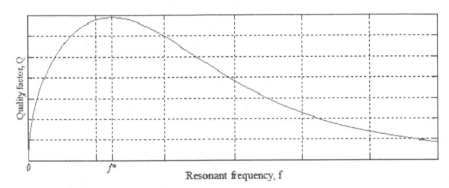

Fig. 17. Quality factor of a coil versus frequency.

There is one more way to increase the quality factor of the coils. Note that, from (26) quality factor of a coil is $Q = \dfrac{\omega L}{R}$ and following (90) and (93) both L and R are frequency dependent variables. So it is possible to find such frequency of operation f that maximize the quality factor (the typical dependency is shown in figure 17). It is important to mention, that since analysis is limited to external capacitance case, this optimal, in the sense of the quality factor maximization, frequency can be obtained by adjusting the external capacitance value. Resonant frequency in presence of the external capacitance can be expressed as (95).

$$f = \frac{1}{2\pi\sqrt{L(C + C_{ext})}} \tag{95}$$

It is worth to mention, that (95) tends to be self resonant frequency, in the case that C_{ext} is null.

Due to the quite complex form of (90) and (95) it is hard to get an analytical expression for the optimal resonant frequency (or the optimal external capacitance for the particular coil). But it is possible to plot Q versus f and from this curve the value of frequency that maximize the quality factor can be identified. After evaluating the optimal frequency of operation and substituting it to (95) optimal value for the external capacitance can be obtained.

3.2 Parameters to increase efficiency of overall power transfer system
As discussed at the beginning of this part figure of merit U is proportional to the mutual inductance M, square roots of coils' quality factors, and inverse proportional to the coils'

inductances. Previous section presents detailed discussion how to maximize the quality factors of coils. Here considerations for the efficiency maximization of all system will be presented (according to (68) when optimal power transmission considerations are satisfied and when U-maximization is equal to overall efficiency maximization).

In case when both antennas have the same axis of cylindrical symmetry and when the condition $Diam << \lambda$, where λ is the wavelength, is satisfied, the mutual inductance could be calculated as (96).

$$M \approx \frac{\pi}{2} \frac{\mu_0 \left(\dfrac{Diam_1}{2} \dfrac{Diam_2}{2} \right)^2 n_1 n_2}{d^3} \tag{96}$$

where subscript $1,2$ stands for source and load coils respectively and d for the distance between the centers of antennas. Substituting (90),(93) and (96) into (81) make it possible to identify critical parameters for efficiency maximization.

Following Ref.[3] U can be maximized by:

1. Decreasing the resistivity, ρ, of coil wires. This normally implies usage of copper or silver wires. Note, here that due to the skin effect, the highest current density is near the edge of a wire, so fully silver wires are not necessary, but just thin cover with silver coating is enough. Also extremely low temperatures of operation together with the superconducting materials could provide nearly perfect performance[2].
2. Increase of wire radius, a, which decreases of resistance. And the efficiency on optimal frequency increases. Note that in typical application wire radius can't be increased, due to space limitations.
3. Increase of coil radius, r, which increases the efficiency with respect to constant distance d. But the coil radius suffers even more from the space limitations than the wire radius.
4. Increase of number of turn of coil, n, leads to the coupling amplification.

In the preceding efficiency maximization discussion it was assumed that optimal conditions (73) for the power transfer are always fulfilled i.e.

$$D_1 = D_2 = D_0^* = 0$$
$$U_0^* = \sqrt{1 + U^2} \tag{73}$$

In section 2.4.1 the case when the second condition is not valid was discussed. But there were no explanation of $D_1 \neq D_2$ case. Note that the meaning of this mismatch is the non-equality of coils resonant frequencies. In figure 18 the influence of resonant frequency detuning on efficiency of system is presented. Calculations were held for a virtual system with $Q=300$.

In figure 18 it can be seen that frequency detuning has a critical effect on lowering the efficiency, especially on longer distances of power transfer. This effect will even be intensified with the growth of quality factor. This observation leads that the tuning of coils to be very important issue in the point of view of system efficiency.

---

[2] Note that both superconducting conditions and silver coating are quite expensive solutions and can't be considered as real efficiency enlargement methods.

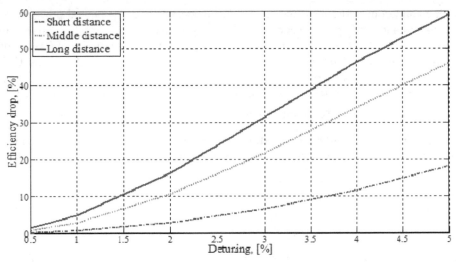

Fig. 18. Influence of coils according to the resonance frequencies detuning[3].

3.3 Different connection ways

In the presented discussion it was assumed that the source coil gets power from some high frequency generator, but the way of such connection was never discussed. Also according to (95) it is assumed that the external and the internal capacitance connected in parallel without providing any theoretical background for this. This section is going to briefly cure these shortcomings.

As shown in Ref.[7], if generator is directly connected to the center of antenna, the source coil could be described as an open end transmission line with length equal to quarter-wave and consequently series connected equivalent circuit[4]. In figure 19 it is described.

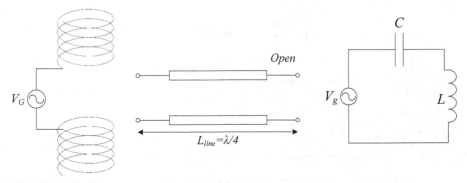

Fig. 19. Open end helical antenna model.

---

[3] On this figure "short", "middle" and "long" distances terms has no specific meaning, only used for just three different points were assumed and influence of frequency detuning on this particular distances were studied.

[4] The discussion in this section is held only for the generator and the source coil, but the same is strictly applicable for the load and the load coil.

At such a connection series resonance occurs (i.e. resonance of currents) and power transfer is possible, due to the large magnetic field which is produced by current. In contrary, if the generator is connected to the antenna edge (refer figure 19), the description of source coil is equivalent to short end transmission line which is depicted by parallel connection of inductance and self capacitance in lumped parameters' case. Obviously that for such an arrangement resonance of voltages will appear, i.e. input impedance of antenna from the generator side will tend to be infinity when frequency of operation approaches to $\dfrac{1}{2\pi\sqrt{LC}}$ (self resonant frequency of a coil). And, it can be noted that a short end helical antenna connection makes efficient wireless power transfer be impossible.

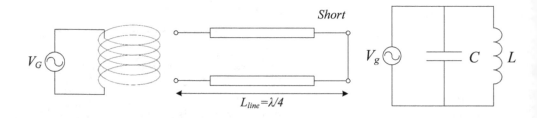

Fig. 20. Short end helical antenna model.

However, by adding an external capacitor to the coil some changes occur. With respect to generator and coil the external capacitance could be connected in four possible ways: in series or parallel to the open end antenna and again in series or in parallel to the short end antenna. Following the discussion in Ref.[7] the results of different connection types are summarized in Table 3.

Connection	Resonant frequency	Efficient wireless power transfer
Series to open end	Up	Possible
Parallel to open end	No change	Impossible
Series to short end	Down	Possible
Parallel to short end	Down	Impossible

Table 3. Influence of different types of external capacitor connections.

From table 3 the efficient power transfer is possible only in case of series connection of external capacitor to the open end antenna or in case of parallel connection to the short end antenna. However, in the view point of the reduction of the operation frequency the series connection of capacitor to the short end antenna would be preferred. In lumped parameters this connection is presented in figure 21 (subscript 1 refers to the sours coil). In figure 2.21 R_{ac} and R_G represents the coil loss and output generator resistance respectively (in case of load coil R_L will appear instead of R_G)

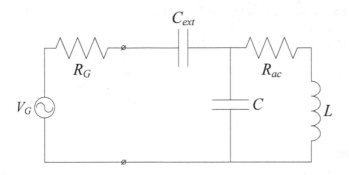

Fig. 21. Series connection of external capacitor to the short end antenna.

However, such type of connection has one important drawback in the sense of overall wireless power transfer system efficiency maximization. From (73) the second condition for the efficient power transfer is $U_1 = U_2 = U_0^* = \sqrt{1+U^2}$ and according to figure 21 coefficients U_1 and U_2 are derived as (97).

$$U_1 = \frac{R_G}{2L_1\Gamma_1}, \ U_2 = \frac{R_L}{2L_2\Gamma_2} \tag{97}$$

Also it can be noted

$$U \propto d \tag{98}$$

Equation (97) together with (73) leads to conclusion that in case of mismatching of generator's output resistance, R_G, is not equal to load resistance, R_L, there is no way to equalize U_1 and U_2. Moreover it is impossible to satisfy the optimal condition by (59) because U is proportional to distance, d, between the coils, as (98). Consequently to apply such a system to wireless power transfer it should be guaranted that generator's resistance is equal to the load one and operation occurs only for one predetermined distance. Otherwise, only semi-optimal operation is possible.

To overcome the restriction of constant wireless power transfer distance and $R_L=R_G$ requirements it is possible to use indirect way of feeding of the resonant coils. If the generator is connected to one turn of wire (feed[5] coil) which is inductively coupled to the source coil, then a degree of freedom can be obtained. By adjusting feeding coil position with respect to the source coil, U_1 can be tuned to get optimal value corresponding to (73).

4. Conclusion

In this chapter CMT is used for the system analysis. It has several useful features, and the most important one is that the wireless power transfer can be presented as a set of first order linear differential equations which could be further analyzed by means of the control theory. While, the application of CMT to the wireless power transfer system are based on several assumptions; at the first, loss in system must be small enough to consider it as the

---

[5] One turn of coil, connected to the load will be called "drag coil".

perturbation, and at the second, the field profile can be calculated by the superposition of each resonant object influence.

From the system analysis it is shown that series resonance is the necessary condition for the efficient power transfer and this efficiency could be maximized by increasing the cross-sectional area of the antenna wire, coil radius, or conductivity of the wire material. At the same time theoretical prediction of an optimal external capacitance by standard means are demonstrated to be impossible. Also in this chapter optimal power transfer system's conditions are derived.

5. References

[1] S. J. Orfanidis "Electromagnetic waves and antennas", ECE Department Rutgers University, NJ 08854-8058, 2008.

[2] H.A. Haus, Waves and Fields in Optoelectronics, Prentice-Hall, New Jercey, 1984.

[3] Karalis, R.E. Hamam, J.D. Joannopoulos, M. Soljacic U.S. patent US 2009/0284083 A1, 2009.

[4] A. Kurs "Power transfer through strongly coupled resonances", MIT, master thesis, 2007.

[5] R.E. Hamam, A. Karalis, J.D. Joannopoulos, M. Soljacic "Coupled-mode theory for general free-space resonant scattering of waves", Physical review A, vol. 75, issue 5, ID 053801, 2007.

[6] T. Imura, H. Okabe, Y. Hori "Basic experimental study on helical antennas of wireless power transfer for electric vehicles by using magnetic resonant couplings", VPPC, pp. 936 – 940, Dearborn, MI, 2009.

[7] T. Imura, H. Okabe, T. Uchida, Y. Hori "Study on open and short end helical antennas of wireless power transfer using magnetic resonant couplings", The 10th University of Tokyo – Seoul National University Joint Seminar on Electric Engineering, pp. 175-180, Seoul, 2010.

[8] D.W. Knight http://www.g3ynh.info/zdocs/magnetics

[9] H. Nagaoka "The inductance coefficient of solenoids", Journal of the College of Science, Imperial University, vol. XXVII, article 6, Tokyo, 1909.

[10] E. B. Rosa "Formulas and tables for the calculation of mutual and self-inductance", Scientific Papers, 1 p., I, 237 p.,1916.

[11] R. G. Medhurst "H.F. resistance and self-capacitance of single-layer solenoids", Wireless Eng., vol. 24, pp. 35–43, Feb. 1947.

3

The Phenomenon of Wireless Energy Transfer: Experiments and Philosophy

Héctor Vázquez-Leal, Agustín Gallardo-Del-Angel,
Roberto Castañeda-Sheissa
and Francisco Javier González-Martínez
University of Veracruz
Electronic Instrumentation and Atmospheric Sciences School
México

1. Introduction

There is a basic law in thermodynamics; the law of conservation of energy, which states that *energy may neither be created nor destroyed just can be transformed*. Nature is an expert using this physics fundamental law favouring life and evolution of species all around the planet, it can be said that we are accustomed to live under this law that we do not pay attention to its existence and how it influence our lives.

Since the origin of the human kind, man has been using nature's energy in his benefit. When the fire was discovered by man, the first thing he tried was to transfer it where found to his shelter. Later on, man learned to gather and transport fuels like mineral charcoal, vegetable charcoal, among others, which then would be transformed into heat or light. In fact, energy transportation became so important for developing communities that when the electrical energy was invented, the biggest and sophisticated energy network ever known by the human kind was quickly built, that is, the electrical grid. Such distribution grid pushed great advances in science oriented to optimize the efficiency on driving such energy. Nevertheless, is common to lose around 30% of energy due to several reasons. Nowadays, there are some daily life applications that could use an energy transport form without cables, some of them could be:

- Medical implants. The advance in biomedical science has allowed to create biomedical implants like: pacemakers, cochlear implants, subcutaneous drug supplier, among others.

- Charge mobile devices, electrical cars, unmanned aircraft, to name a few.

- Home appliances like irons, vacuum cleaners, televisions, etc.

Such potential applications promote the interest to use a wireless energy transfer. Nevertheless, nature has always been a step beyond us, doing energy distribution and transformation since a long while without the need of copper cables. The biggest wireless transfer source known is solar energy; nature uses sunlight to drive the photosynthesis process, generating this way nutrients that later on will become the motor for the food chain and life. At present, several ways to turn sunlight into electrical power have been invented, among them, the photo-voltaic cells are the most popular. However, collecting solar energy is just the first step, the distribution of this energy is the other part of the problem, that is, the new objective is to wirelessly transfer point to point the energy.

This new technological tendency towards wireless energy is not as new as one might think. It is already known that the true inventor of the radio was Nikola Tesla, therefore, makes sense to think this scientist inferred, if it was possible to transfer information using an electromagnetic field, it would be also possible to transfer power using the same transmission medium or vice versa. Thus, in the early 19-nth century this prominent inventor and scientist performed experiments (Tesla (1914)) regarding the wireless energy transfer achieving astonishing results by his age. It has been said that Tesla's experiments achieved to light lamps several kilometres away. Nevertheless, due to the dangerous nature of the experiments, low efficiency on power transfer, and mainly by the depletion of financial resources, Tesla abandoned experimentation, leaving his legacy in the form of a patent that was never commercially exploited.

Electromagnetic radiation has been typically used for the wireless transmission of information. However, information travels on electromagnetic waves which are a form of energy. Therefore, in theory it is possible to transmit energy similarly like the used to transfer information (voice and data). In particular, it is possible to transfer in a directional way great powers using microwaves (Glaser (1973)). Although the method is efficient, it has disadvantages: requires a line of sight and it is a dangerous mechanism for living beings. Thus, the wireless energy transfer using the phenomenon of electromagnetic resonance has become in a viable option, at least for short distances, since it has high efficiency for power transfer. The authors of (Karalis et al. (2008); Kurs (2007)) claim that resonant coupling do not affect human health.

At present, energy has been transferred wirelessly using such diverse physical mechanisms like:

- Laser. The laser beam is coherent light beam capable to transport very high energies, this makes it in an efficient mechanism to send energy point to point in a line of sight. NASA (NASA (2003)) introduced in 2003 a remote-controlled aircraft wirelessly energized by a laser beam and a photovoltaic cell infra-red sensitive acting as the energy collector. In fact, NASA is proposing such scheme to power satellites and wireless energy transfer where none other mechanism is viable (NASA (2003)).

- Piezoelectric principle (Hu et al. (2008)). It has been demonstrated the feasibility to wirelessly transfer energy using piezoelectric transducers capable to emit and collect vibratory waves.

- Radio waves and Microwaves. In (Glaser (1973)) is shown how to transmit high power energy through long distances using Microwaves. Also, there is a whole research field for rectennas (J. A. G. Akkermans & Visser (2005); Mohammod Ali & Dougal (2005); Ren & Chang (2006); Shams & Ali (2007)) which are antennas capable to collect energy from radio waves.

- Inductive coupling (Basset et al. (2007); Gao (2007); Low et al. (2009); Mansor et al. (2008)). The inductive coupling works under the resonant coupling effect between coils of two LC circuits. The maximum efficiency is only achieved when transmitter and receiver are placed very close from each other.

- "Strong" electromagnetic resonance. In (Karalis et al. (2008); Kurs (2007)) was introduced the method of wireless energy transfer, which use the "strong" electromagnetic resonance phenomenon, achieving energy transfer efficiently at several dozens of centimetres.

Transferring great quantities of power using magnetic field creates, inevitably, unrest about the harmful effects that it could cause to human health. Therefore, the next section will address this concern.

2. Electromagnetic waves and health

Since the discovering of electromagnetic waves a technological race began to take advantage of transferring information wirelessly. This technological race started with Morse code transmission, but quickly came radio, television, cellular phones and the digital versions for all the mentioned previously. Adding to the mentioned before, in the last decade arrived an endless amount of mobile devices capable to communicate wirelessly; these kind of devices are used massively around the globe. As a result, it is common that an average person is subjected to magnetic fields in frequencies going from Megahertz up to the Gigahertz. Therefore, the concerns of the population about health effects due to be exposed to all the electromagnetic radiation generated by our society every day. Besides, added to the debate, is the concern for the wireless energy transfer mechanisms working with electromagnetic signals.

Several studies have been completed (Breckenkamp et al. (2009); Habash et al. (2009)) about the effects of electromagnetic waves, in particular for cellular phones, verifying that just at the upper international security levels some effects to genes are noticed. In (Peter A. Valberg & Repacholi (2007)) is assured that it is not yet possible to determine health effects either on short or long terms due by the exposition to electromagnetic waves like the ones emitted by broadcasting stations and cellular networks. Nevertheless, in (Valborg Baste & Moe (2008)) a study was performed to 10,497 marines from the Royal Norwegian Navy; the result for the ones who worked within 10 meters of broadcasting stations or radars, was an increase on infertility and a higher birth rate of women than men. This increase of infertility agrees with other study (Irgens A & M (1999)) that determined that the semen quality decay in men which by employment reasons (electricians, welders, technicians, etc.) are exposed to constant electromagnetic radiation including microwaves. These studies conclude that some effects on the human being, in fact occur, mainly at high frequencies.

3. Acoustic and electrical resonance

The mechanical resonance or acoustic is well known on physics and consists in applying to an object a vibratory periodic action with a vibratory period that match the maximum absorption energy rate of the object. That frequency is known as resonant frequency. This effect may be destructive for some rigid materials like when a glass breaks when a tenor sings or, in extreme cases, even a bridge or a building may collapse due to resonance; whether it is caused by the wind or an earthquake.

Resonance is a well known phenomenon in mechanics but it is also present in electricity; is known as electrical resonance or inductive resonance. Such phenomenon can be used to transfer wireless energy with two main advantages: maximum absorption rate is guaranteed and it can work in low frequencies (less dangerous to humans). When two objects have the same resonant frequency, they can be coupled in a resonant way causing one object to transfer energy (in an efficient way) to the other. This principle can be exploited to transmit energy from one point to another by means of an electromagnetic field. Next, three wireless energy transfer mechanisms are described:

1. **Inductive coupling (Mansor et al. (2008))** is a resonant coupling that takes place between coils of two LC circuits with the same resonant frequency, transferring energy from one coil to the other as it can be seen in figure 1(a). The disadvantage of this technique is that efficiency is lost as fast as coils are separated.

2. **Self resonant coupling (Karalis et al. (2008))**. The self resonance occur in a natural way for all coils (L), although the frequency

$$f_r = 1/(2\pi\sqrt{LC_p}),$$

is usually too high because the parasitic capacitance (C_p) value is too low. Nevertheless, in (Karalis et al. (2008)) was shown that it is possible to achieve good efficiency with a scheme like the one shown in figure 1(b). For the coupling to surpass the 40% reported in Karalis et al. (2008) the radius (r) for the coil must be much lower than wavelength (λ) of the resonant frequency and the optimum separation (d) for a good coupling should be such $r \ll d \ll \lambda$, in such a way that the coupling is proportional to (Urzhumov & Smith (2011))

$$\frac{r}{\lambda}\frac{r^3}{d^3}.$$

There are two fundamental differences for the simple inductive coupling in figure 1(a), those are: the capacitance of the LC circuit is parasitic, not discrete, and now coils (T y R) are coupled to two one spire coils L_S and L_L, those act as the emitter source and receiver coils, respectively. The coil's self resonant frequency depends of its parasitic capacitance, that is the reason the frequency is very high (around the GHz range). Therefore, to achieve lower self resonance frequency (< 10Mhz) it is necessary to use thick and spaced copper wire to create higher parasitic capacitance, reducing the self resonance frequency down to the megahertz range. In fact, in Karalis et al. (2008) and Kurs (2007) is reported an experiment using cable with 3 cm. radius. The efficiency on the power transfer with respect to the distance has an inverse relationship to the radius of the coil, that is why the experiments reported in Karalis et al. (2008) and Kurs (2007) coils have 30 cm. radius.

3. In figure 1(c), the coupling scheme shown can be named as **modified resonant inductive coupling**, this is a modification for the strong resonant coupling (see figure 1(b)). The modification consists in exchange the parasitic capacitance C_p for a discrete capacitance C. Thus, the need for large and thick cable is eliminated.

4. Experimentation

Triangular and circular coils are going to be employed in order to establish an inductive resonant coupling as shown in figure 1(a) and figure 3.

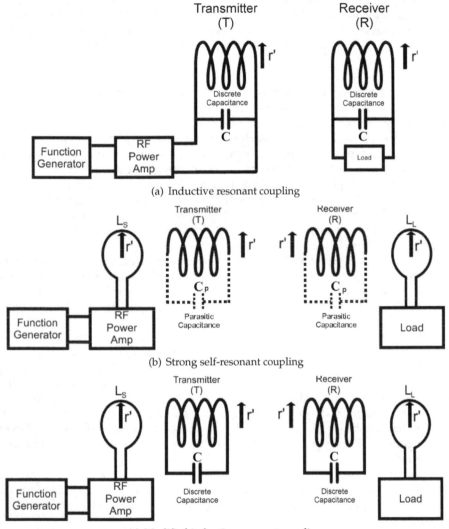

(a) Inductive resonant coupling

(b) Strong self-resonant coupling

(c) Modified inductive resonant coupling

Fig. 1. Coupling schemes

4.1 Inductive resonant coupling at low frequency

This experiment was designed (J.A. Ricaño-Herrera et al. (2010)) to visualize the radiation pattern and the efficiency of an inductive resonant coupling. First, the generating coil was kept in a fixed position while the receiver coil (R) revolves around the generating coil (T), at a fixed distance and with constant angular displacement completing 360 degrees (see figure 2(a)). The experiment shows that the produced energy by the transmitter coil T propagates at $90°$ in front of the generating coil and at $90°$ behind the same coil. In another stage of the experiment, two coils were placed in parallel and concentric at a distance of zero centimeters, then they were moved away. The results are shown in figure 2(b). It can be seen from figure

2(c) that the maximum efficiency for voltage gain is around the 50% (at zero centimeters). The result is logical after observing the radiation pattern shown in figure 2(b), because a radiation back lobe is wasted. Figure 2(c) shows, beyond the 8 cm. distance, the voltage gain for the system falls below the 5% value. The back lobe could be reused using a reflecting surface for the magnetic field.

4.2 Comparison between circular and triangular coils at medium frequencies

The phenomenon is well known in mechanics is also present in electricity and is called electrical resonance or inductive resonance

Differences between circular and triangular coils are related to the geometry of the coil, frequency response and radiation pattern. However, these differences produce similar results. The difference in the geometry of coils cause subtle changes in the inductance altering its resonant frequency. Figure 4 was obtained by a S parameter analyser showing several resonant frequencies for the circular and triangular coils. From this figure, the first resonant frequencies can be observed in the range of 21 MHz to 26 MHz for both kind of coils. It is important to recall that just two circular or two triangular coils were used in all experiments to complete the system (figure 1(a) and figure 3).

To determine the working frequency, each pair of coils were tested with a RF generator and a spectrum analyser. Figure 5 shows the frequency sweep for circular coils and figure 6 presents the frequency sweep for triangular coils. The working frequency can be observed between 21 MHz and 27 MHz. This range of frequencies is due to imperfections of coils and not being identical.

Once working frequencies were found for each pair of (circular and triangular) coils, we are ready to initiate the energy transfer experiment. In this experiment, the RF generator was connected to one circular or triangular coil (called Transmitter coil); the spectrum analyser was connected to the other circular or triangular coil (called Receiver coil). Initially, both coils were separated 0 centimetres. After that, one coil was displaced up to 25 centimetres in steps of 1 centimetre. Figure 7 shows the received power for circular and triangular coils in the range from 0 centimetres to 25 centimetres. The frequency distribution (for four distances) is shown in figure 8. In this figure it can be observed that the amplitude, bandwidth and spectral density decrease.

The normalized efficiency for the receiver coil was calculated considering that the maximum power will be always close to 0 centimetres. In this scheme, the efficiency is proportional to the received power (see figure 7 and figure 9). Figure 9 shows the efficiency for circular and triangular coils.

From figures 4, 5, and 6 can be seen that for the same coil system (circular or triangular), several resonance frequencies were obtained, which can be used to transfer power efficiently simultaneously.

The graphical form of the spatial distribution of energy was measured. The radiation pattern for circular and triangular coils is shown in figure 10. It is interesting to observe that, at low frequency figure 7 shows the efficiency decaying as distance increases. This can be explained observing figure 10, it shows for both cases a deformed radiation pattern with respect to the low frequency pattern (see figure 2(b)); at low frequencies the radiation pattern is uniform and has 2 lobes centred on the x axis. Nevertheless, at medium frequencies, the radiation pattern is deformed and has four lobes not centred at x axis. Such lobes are uniformly spaced at 90° from each other, starting at 45° from x axis. This phenomenon explains the fast decay of the

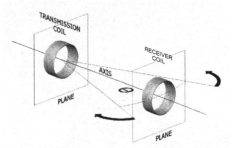

(a) Experimental process for the revolving coil

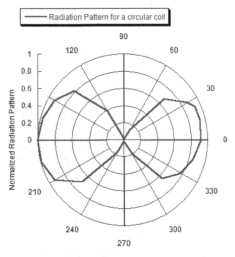

(b) Radiation pattern

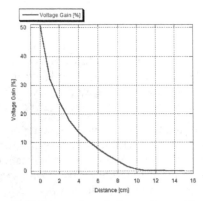

(c) Voltage gain of the system with respect to the voltage ratio

Fig. 2. Generator coil radiation pattern at low frequency (1.4 MHz).

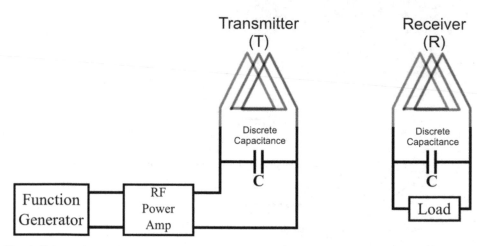

Fig. 3. Triangular coil experiment.

efficiency with respect to the distance shown in figure 7, since coils where placed in a coaxial location, this induced an inefficient coupling, which can be improved turning the emitter coil (A) −45° to make one lobe match the coaxial axis. Also, it is necessary a future research of the radiation pattern shape at higher frequencies, since this work showed the radiation pattern varies as frequency changes for the inductive resonant coupling to achieve more efficiency.

This work showed experiments with the coupling scheme shown in figure 1(a) (two coils), which is different from the scheme shown in figure 1(c) (four coils); in scheme 1(c), the single coil L_s generates a radiation pattern, which is coupled to coil T. Coils T and R in this scheme work as lenses concentrating the energy and improving directivity from coil L_s to coil L_l. Analysis at different frequencies of the radiation pattern could show changes in directivity and the existence of several useful resonating frequencies for the strong coupling resonance (figure 1(c)).

5. Philosophy

The wireless energy with: high power (>100W), reaching longer distance (>10m), having good efficiency (>70%), without health concerns, and low cost is a dream that keeps the attention of researchers around the planet. Nevertheless, in order to make a dream come true it is necessary innovating ideas or even radical ones, to provide the answer for the big questions opposed to achieve the goal. Therefore, innovation is required on the following directions:

- Coils with different geometries. Coils employed on the reported experiments with spirals are circular and triangular. Nevertheless, new geometries (hexagonal, multiform) can be used and thus modify the radiation pattern, this modification on the pattern seeks the increase of: directionality, distance and/or efficiency. With completely different coils, like hexagonal, multiform, highly non-linear radiation patterns could be generated like the ones shown in figure 11.

- Using new materials to improve efficiency. For instance, from the self-resonance coils in experiment 2, it can be achieved using a coil-capacitor device. This coil could be designed

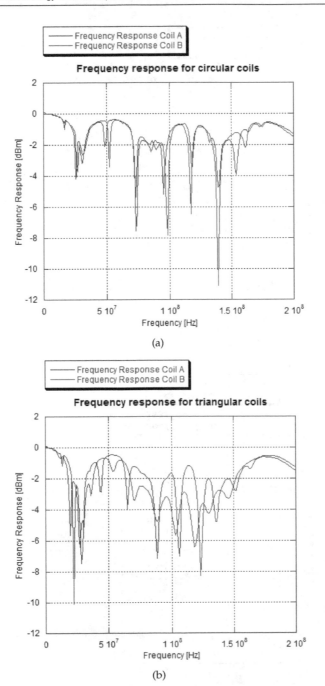

Fig. 4. Resonant frequencies for (a) circular and (b) triangular coils for frequencies lower than 200 MHz.

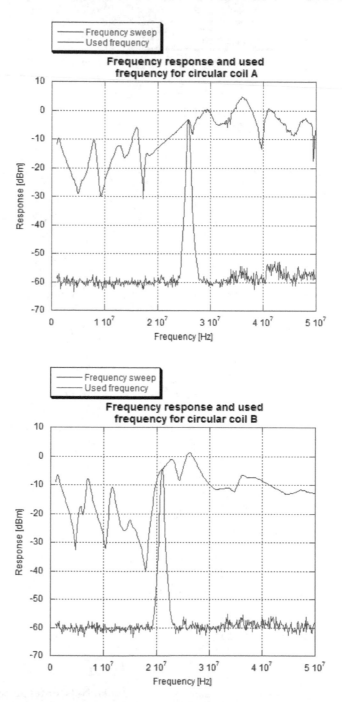

Fig. 5. Frequency sweep and working frequency for a pair of circular coils.

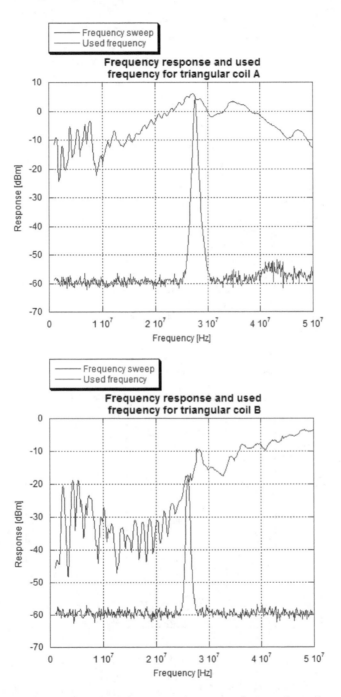

Fig. 6. Frequency sweep and working frequency for a pair of triangular coils.

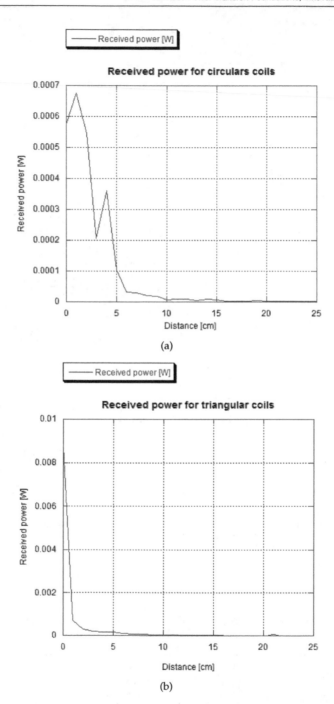

Fig. 7. Received power for (a) circular and (b) triangular coils.

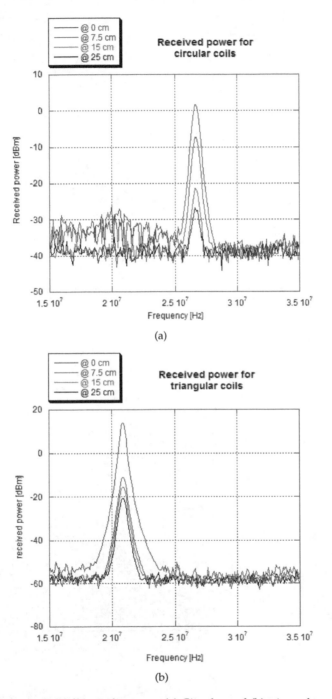

Fig. 8. Received power for different distances. (a) Circular and (b) triangular coils.

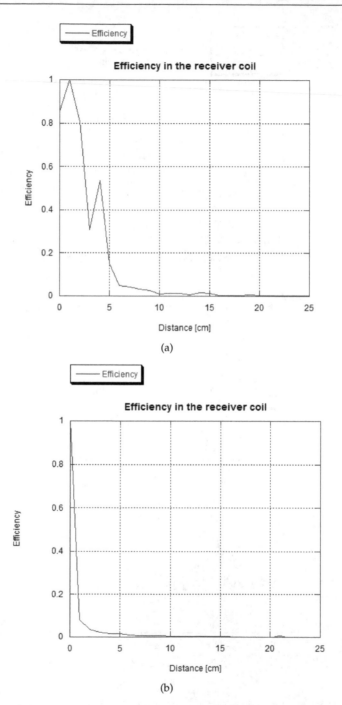

Fig. 9. Normalized efficiency for (a) circular and (b) triangular coils.

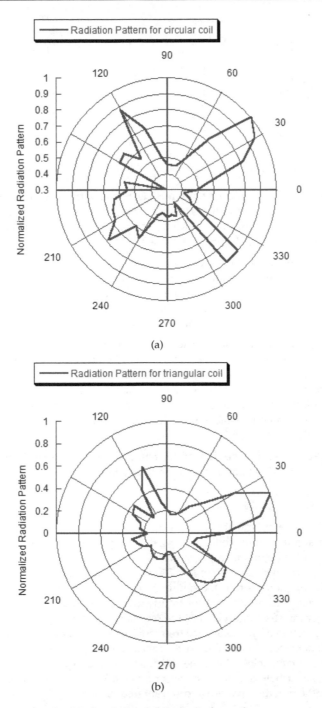

Fig. 10. Radiation pattern for (a) circular and (b) triangular coils.

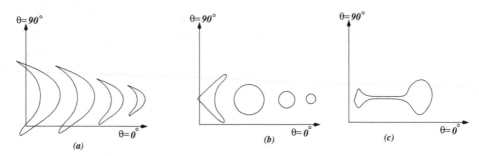

Fig. 11. New radiation patterns.

in such a way that between each spire a dielectric material is placed to create parasitic capacitance along all the coil spires. Therefore, parasitic capacitance will be big enough to achieve self-resonance on the order of MHz. The advantage of this coil-capacitor is that no longer thick coils and spaced spires will be needed.

- In (Urzhumov & Smith (2011) was demonstrated that using metamaterials could improve performance of coupled resonant systems in near field. They proposed a power relay system based on a near-field metamaterial superlens. This is the first step toward optimization of the resonant coupling phenomenon in near field, the next will be the design of coils implemented with metamaterials looking to affect directionality or efficiency.

- Inductive coupled multi-resonant systems. Using amorphous or multiform coils could generate multiple resonant frequencies that could be employed in the transfer of energy using more than one resonant frequency, this will depend on their emitting pattern and efficiency extent. Another possible application for multi-resonant systems is transmission of energy and information a the same time using different channels. For instance, using the information channel to establish the permission for the energy transfer and features like power levels.

- A waveguide designed for the transmitter coil and a reflecting stage in order to use the back lobe of the radiation pattern, may help to improve efficiency of the power transfer.

6. Conclusion

In this work several experiments were performed showing differences and similarities between circular and triangular coils for wireless energy transfer by means of the inductive resonant coupling phenomenon. In particular, showed that the radiation pattern is different for low and middle frequencies. As for low frequencies, two lobes aligned to the x axis were found; for middle frequencies four lobes uniformly spaced but unaligned to the x axis were located. This characteristic deserves deeper study to determine the possibility to use it in order to direct the energy transfer modifying just the resonance frequency. Besides, it was found that the number and position of the resonance frequencies for circular and triangular coils are not similar. This phenomenon could be used to transmit energy or information simultaneously by such resonance frequencies. Also, the efficiency decays exponentially with distance for both geometries, nevertheless, this could be improved taking the advantage of the deforming phenomenon for the radiation pattern at different frequencies.

7. References

Basset, P., Andreas Kaiser, B. L., Collard, D. & Buchaillot, L. (2007). Complete system for wireless powering and remote control of electrostatic actuators by inductive coupling, *IEEE/ASME Transactions on Mechatronics* 12(1).

Breckenkamp, J., Gabriele Berg-Beckhoff, E. M., Schuz, J., Schlehofer, B., Wahrendorft, J. & Blettner, M. (2009). Feasibility of a cohort study on health risks caused by occupational exposure to radiofrequency electromagnetic fields, *BioMed Central Environmental Health* .

Gao, J. (2007). Traveling magnetic field for homogeneous wireless power transmission, *IEEE Transactions on Power Delivery* 22(1).

Glaser, P. E. (1973). Method and apparatus for converting solar radiation to electrical power, *U.S.A Patent* .

Habash, R. W., Elwood, J. M., Krewski, D., Lotz, W. G., McNamee, J. P. & Prato, F. S. (2009). Recent advances in research on radiofrequency fields and health: 2004-2007, *Journal of Toxicology and Environmental Health, Part B* pp. 250–288.

Hu, H., Hu, Y., Chen, C. & Wang, J. (2008). A system of two piezoelectric transducers and a storage circuit for wireless energy transmission through a thin metal wall, *IEEE Transactions on Ultrasonics, Ferroelectrics, and Frequency Control* 55(10).

Irgens A, K. K. & M, U. (1999). The effect of male occupational exposure in infertile couples in norway, *J Occup Environ Med.* 41: 1116–20.

J. A. G. Akkermans, M. C. van Beurden, G. D. & Visser, H. (2005). Analytical models for low-power rectenna design, *IEEE Antennas and Wireless Propagation Letters* 4.

J.A. Ricaño-Herrera, H. Rodríguez-Torres, H. Vázquez-Leal & A. Gallardo-del-Angel (2010). Experiment about wireless energy transfer, *International Congress on instrumentation and Applied Sciences*, CCADET, Cancun, Q.R., Mexico, pp. 1–10.

Karalis, A., Joannopoulos, J. & Soljacic, M. (2008). Efficient wireless non-radiative mid-range energy transfer, *Elsevier Annals of Physics* (323): 34–48.

Kurs, A. (2007). Power transfer through strongly coupled resonances, *Massachusetts Institute of Technology, Master of Science in Physics Thesis* .

Low, Z. N., Chinga, R. A., Tseng, R. & Lin, J. (2009). Design and test of a high-power high-efficiency loosely coupled planar wireless power transfer system, *IEEE Transactions on Industrial Electronics* 56(5).

Mansor, H., Halim, M., Mashor, M. & Rahim, M. (2008). Application on wireless power transmission for biomedical implantable organ, *Springer-Verlag Biomed 2008 proceedings* 21 pp. 40–43.

Mohammod Ali, G. Y. & Dougal, R. (2005). A new circularly polarized rectenna for wireless power transmission and data communication, *IEEE Antennas and Wireless Propagation Letters* 4.

NASA (2003). Beamed laser power for uavs, *Dryden Flight Research Center* .

Peter A. Valberg, T. E. v. D. & Repacholi, M. H. (2007). Workgroup report: Base stations and wireless network radiofrequency (rf) exposures and health consequences, *Environmental Health Perspectives* 115(3).

Ren, Y.-J. & Chang, K. (2006). 5.8-ghz circularly polarized dual-diode rectenna and rectenna array for microwave power transmission, *IEEE Transactions on Microwave Theory and Techniques* 54(4).

Shams, K. M. Z. & Ali, M. (2007). Wireless power transmission to a buried sensor in concrete, *IEEE Sensors Journal* 7(12).

Tesla, N. (1914). Apparatus for transmitting electrical energy, *USA Patent 1119732* .

Urzhumov, Y. & Smith, D. R. (2011). Metamaterial-enhanced coupling between magnetic dipoles for efficient wireless power transfer, *Phys. Rev. B* 83(20): 205114–10.

Valborg Baste, T. R. & Moe, B. E. (2008). Radiofrequency electromagnetic fields; male infertility and sex ratio of offspring, *Springer, European Journal of Environmental Epidemiology* pp. 369–377.

4

Magnetically Coupled Resonance Wireless Power Transfer (MR-WPT) with Multiple Self-Resonators

Youngjin Park, Jinwook Kim and Kwan-Ho Kim
Korea Electrotechnology Research Institute (KERI) and
University of Science & Technology (UST)
Republic of Korea

1. Introduction

Wireless power transfer (WPT) has been studied for more than one hundred years since Nikola Tesla proposed his WPT concept. As more and more portable electronic devices and consumer electronics are developed and used, the need for WPT technology will continue to grow.

Recently, WPT via strongly coupled magnetic resonances in the near field has been reported by Kurs et al. (2007). The basic principle of WPT based on magnetically coupled resonance (MR-WPT) is that two self-resonators that have the same resonant frequency can transfer energy efficiently over midrange distances. It was also reported that MR-WPT has several valuable advantages, such as efficient midrange power transfer, non-radiative, and nearly omnidirectional. It is certain that these properties will help to improve the performance of current wireless power transfer systems and be utilized well for various wireless power transfer applications such as electric vehicles, consumer electronics, smart mobile devices, biomedical implants, robots, and so on.

Up to now, several important articles have been published. Karalis et al. (2008) reported detailed physical phenomena of efficient wireless non-radiative mid-range energy transfer. Sample et al. (2010) reported an equivalent model and analysis of an MR-WPT system using circuit theory, and Hamam et al. (2009) introduced an MR-WPT system that used an intermediate self-resonator coil to extend the coverage of wireless power transfer that is coaxially arranged with both Tx and Rx self-resonant coils.

In Figure 1, a practical application model of wireless power charging of multiple portable electronic devices using MR-WPT technology is illustrated. Multiple devices are placed on the Rx self-resonator, which is built into the desk, and the Tx self-resonator is built into the power plate wall. The Tx self-resonator is strongly coupled with the Rx one and then both Tx and Rx self-resonators transfer energy efficiently even though the Tx self-resonator is geometrically perpendicular to the Rx self-resonator. In order to create this system, it is necessary to characterize power transfer efficiency and especially mutual inductance of the MR-WPT system with two self-resonators arranged perpendicularly. However, there have been few research reports published that analyze the characteristics of MR-WPT regarding a geometrical arrangement between Tx and Rx self-resonators and between Tx or Rx and intermediate self-resonators.

In this article, the characteristics of wireless power transfer between two self-resonators arranged in off-axis positions are reported and the power transfer efficiency of an MR-WPT system with an intermediate self-resonator is analyzed. The intermediate self-resonator is geometrically perpendicular or coaxial to the Tx and Rx self-resonators. To calculate the power transfer efficiency, a modified coupled mode theory (CMT) is applied. In particular, a calculation method and analysis results of mutual inductance between two self-resonators arranged in off-axis positions are presented.

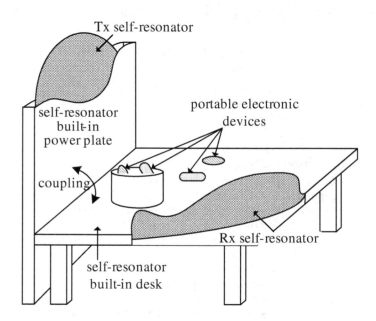

Fig. 1. A practical application of a wireless power transfer system using MR-WPT.

The article is organized as follows. In Section 2, the configuration and modeling of an MR-WPT system with an intermediate self-resonator is illustrated and the power transfer efficiency of the system is derived. In Section 3, mutual inductance between two self-resonators is derived for rectangular and circular coils. In Section 4, two MR-WPT systems with intermediate self-resonators are fabricated and formula derivation, analysis results, and design procedures are verified by experimental measurement.

2. Illustration and modeling of an MR-WPT system with an intermediate self-resonator

Figure 2 shows the configuration of an MR-WPT system with multiple self-resonators. It consists of three self-resonators (Tx, Rx, and intermediate), a source coil, and a load coil. The centers of the Tx, Rx, and intermediate self-resonators are $(0, 0, -D_{1m})$, $(0, 0, D_{2m})$, and $(-D_h, 0, 0)$, respectively. Each coil is loaded with a series of high Q capacitors in order to adjust the target resonant frequency and prevent change of the resonant frequency due to unknown objects. It should be noted that the intermediate resonant coil is arranged perpendicularly with both Tx and Rx self-resonators. By referring to Haus (1984) and Hamam et al. (2009), a

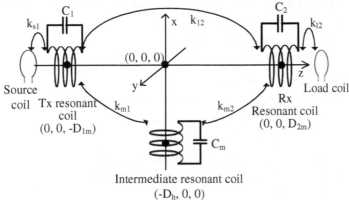

Fig. 2. Configuration of an MR-WPT system with an intermediate self-resonator.

modified CMT equation in matrix form can be written as:

$$
\begin{pmatrix} \frac{d}{dt}a_1(t) \\ \frac{d}{dt}a_m(t) \\ \frac{d}{dt}a_2(t) \\ S_{-1} \\ S_{-2} \end{pmatrix} = \begin{pmatrix} -(i\omega_1 + \Gamma_1) - k_{s1} & ik_{1m} & ik_{12} & \sqrt{2k_{s1}} & 0 \\ iM_{m1} & -(i\omega_m + \Gamma_m) & ik_{m2} & 0 & 0 \\ ik_{12} & ik_{2m} & -(i\omega_2 + \Gamma_2) - k_{l2} & 0 & 0 \\ \sqrt{2k_{s1}} & 0 & 0 & -1 & 0 \\ 0 & 0 & \sqrt{2k_{l2}} & 0 & 0 \end{pmatrix} \begin{pmatrix} a_1(t) \\ a_m(t) \\ a_2(t) \\ S_{+1} \\ S_{+2} \end{pmatrix}.
$$

(1)

k_{s1}, k_{l2}, k_{m1}, k_{m2}, and k_{12} are coupling coefficients between coils. The parameters are defined as follows:

- $a_i(t)$: mode amplitude of each self-resonator,
- ω_i : angular resonant frequency of each self-resonator, $1/\sqrt{L_iC_i}$,
- Γ_i : intrinsic decay rate of each self-resonator, $R_i/2L_i$,
- L_i and R_i : self-inductance and resistance of each self-resonator,
- C_i: capacitance of each self-resonator (self-capacitance + high-Q capacitor),
- k_{ij} : coupling coefficient between i and j self-resonators, $\omega M_{ij}/2\sqrt{L_iL_j}$,
- M_{ij} : mutual inductance between i and j self-resonators,
- $S_{\pm 1}$: field amplitude of an incident field (+) and a reflected field (−) at the source,
- $S_{\pm 2}$: field amplitude of an incident field (+) and a reflected field (−) at the load,
- i and $j(= 1, 2, m, s, l)$: 1 (Tx self-resonator), 2 (Rx self-resonator), m (intermediate self-resonator), s (source coil), l (load coil).

To simplify Equation 1, it is assumed that $k_{12} \approx 0$, $\Gamma_1 = \Gamma_2 = \Gamma \neq \Gamma_m$, and $k_{m1} = k_{m2} = k_m$. That is, Tx and Rx self-resonators are identical and the intermediate self-resonator is placed at the center of the Tx and Rx self-resonators ($D_{1m} = D_{2m}$). Also, the coupling coefficient k_m is much higher than k_{12}. Then, the transmission coefficient (S_{21}) from source to load is obtained as:

$$S_{21} = \frac{-2U_m^2 U_0}{[1 + U_0 - iX_1][1 + U_0 - iX_2][1 - iX_m] + U_m^2[2(1 + U_0) - i(X_1 + X_2)]}, \tag{2}$$

where $X_1 = X_2 = (\omega_{1,2} - \omega_0)/\Gamma$, $X_m = (\omega_m - \omega_0)/\Gamma_m$, $U_0 = k_{s1}/\Gamma = k_{12}/\Gamma$, $U_m = \sqrt{k_m^2/\Gamma\Gamma_m}$, and ω_0 is a target angular resonant frequency. Then, the power transfer efficiency η is obtained as:

$$\eta = |S_{21}|^2. \tag{3}$$

By assuming that the resonant frequency of each self-resonator is the same as the target resonant frequency, that is, $X_1 = X_2 = X_m = 0$, the matching condition for maximum power transfer efficiency in Equation 1 is obtained as:

$$U_0^{opt} = \sqrt{1 + 2U_m^2}.$$

The maximum power transfer efficiency using the condition is rewritten as follows:

$$\eta = \frac{(k_m^2/\Gamma\Gamma_m)^2}{(\sqrt{1 + 2k_m^2/\Gamma\Gamma_m} + 1 + k_m^2/\Gamma\Gamma_m)^2} = \frac{U_m^4}{(\sqrt{1 + 2U_m^2} + 1 + U_m^2)^2}. \tag{4}$$

To calculate Equation 4, three unknown parameters of k_m, Γ, and Γ_m should be determined. The intrinsic decay rates of Γ and Γ_m are determined by the resistance and inductance of each self-resonator. k_m is calculated by mutual inductance between two self-resonators and the self-inductance of each self-resonator. It can also be noted that proper matching in an MR-WPT system with k_m fixed can be accomplished by varying k_{12} for maximum power transfer.

In the next section, the calculation method of mutual inductance is presented for the case of circular and rectangular types of self-resonators.

3. Derivation of mutual inductance

3.1 Mutual inductance between two circular self-resonators
3.1.1 Configuration and derivation
In Figure 3, two circular self-resonators are arranged coaxially and perpendicularly. D is the distance between two coils. For the calculation of mutual inductance, it is assumed that the coils are filamentary and current is uniformly distributed on the coils. Mutual inductance M_{12} between two coils is written as:

$$M_{12} = \frac{N_1 N_2}{I} \int_{S_2} \vec{B} \cdot d\vec{s}_2. \tag{5}$$

By referring to Good (2001), the magnetic flux density, $\vec{B}$ at arbitrary spatial points is written as follows:

$$\vec{B} = \hat{\rho} B_\rho + \hat{z} B_z, \tag{6a}$$

where

$$B_z|_{\rho=0} = \frac{\mu_0 I r_1^2}{2(D^2 + r_1^2)^{3/2}},$$

$$B_z|_{\rho\neq0} = \frac{\mu_0 I}{2\pi\rho}\left(\frac{m}{4r_1\rho}\right)^{1/2}\left(\rho K(m) - \frac{r_1 m - (2-m)\rho}{2(1-m)}E(m)\right), \qquad (6b)$$

$$B_\rho|_{\rho\neq0} = \frac{\mu_0 I D}{2\pi\rho}\left(\frac{m}{4r_1\rho}\right)^{1/2}\left(-K(m) + \frac{2-m}{2(1-m)}E(m)\right),$$

with

$$K(m) = \int_0^{\pi/2}(1 - m\cdot\sin^2\theta)^{-1/2}d\theta,$$

$$E(m) = \int_0^{\pi/2}(1 - m\cdot\sin^2\theta)^{1/2}d\theta, \qquad (6c)$$

$$m = \frac{4r_1\rho}{(r_1+\rho)^2 + D^2}.$$

Here, K and E are the complete elliptic integrals of the first and second kinds, respectively. m is the variable of elliptic integrals. N_1 and N_2 are the number of turns of the first and second coils, respectively. In the coaxially arranged system (see Figure 3a), mutual inductance is determined by only z-directed fields (B_z). The magnetic flux density of a circular coil at the points of the same ρ is identical and then the total magnetic flux linkage is obtained by summing the flux of a central circular area and each circular subdivision as well. Therefore, by assuming that N_d is sufficiently large, mutual inductance between two coaxially arranged coils is written as follows:

$$M_{cc} = \frac{2\pi d^2 N_1 N_2}{I}\left\{\frac{B_z|_{n=0}}{4} + \sum_{n=1}^{N_d=r_2/d} n B_z|_n\right\} \approx \frac{2\pi d^2 N_1 N_2}{I}\sum_{n=1}^{N_d=r_2/d} n B_z|_n. \qquad (7)$$

Here, N_d is the total number of subdivisions of the Rx self-resonator.
For the case of two perpendicularly arranged circular coils (see Figure 3b), mutual inductance is determined by only ρ-directed fields. Therefore, the mutual inductance is obtained as follows:

$$M_{pc} = \frac{N_1 N_2}{I}\sum_{n=1}^{N_d=2r_2/d} B_\rho|_n \cdot S_n, \qquad (8)$$

where $S_n = 2nd^2\sqrt{2r_2/nd - 1}$ and is the n-th rectangular area subdivided. For more general cases, see Babic et al. (2010).

3.1.2 Calculation and measurement

For verification of the calculation method, two self-resonators were made as shown in Figures 4 and 10. A target resonant frequency, f_0 was 1.25 MHz. The Tx self-resonator was a helical type ($r = 252$ mm, $H = 90$ mm, $N_1 = 9$ turns, $a = 2.2$ mm). The Rx coil was a spiral type ($r_{in} = 230$ mm, $r_{out} = 300$ mm, $N_2 = 10$ turns, $a = 3.2$ mm). Both of the coils were made of copper pipe ($\sigma = 5.8 \times 10^7$). Using high-Q capacitors, the target resonant frequency of each self-resonator was adjusted. The intrinsic decay rate using measured resistance and inductance, capacitance of high-Q capacitors for each self-resonator, self-inductances, and resonant frequencies of the self-resonators are shown in Table 1. To measure resonant

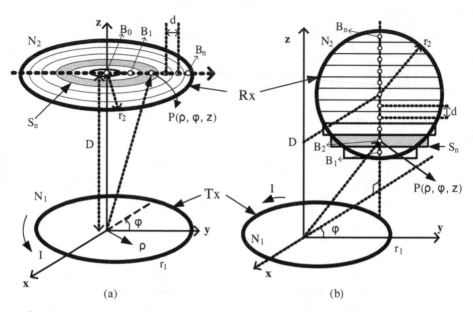

Fig. 3. Configuration of two circular self-resonators for calculation of mutual inductance.

frequency, a vector network analyzer (Agilent 4395A) was used. To measure self-inductance (L) and resistance (R) of each self-resonator, an LCR meter (GWInstek 8110G) was used.

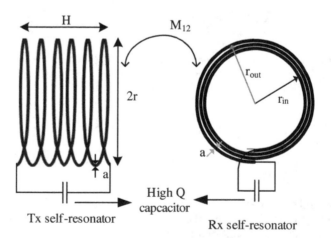

Fig. 4. Schematic drawing of Tx and Rx self-resonators.

To measure mutual inductance, both differential coupling inductance ($L_{m1} = L_1 + L_2 - 2M_{12}$) and cumulative coupling inductance ($L_{m2} = L_1 + L_2 + 2M_{12}$) were measured and then the

	Γ	High-Q capacitor	self-inductance	f_0
Tx (helix)	8168.14	224.40 pF	67.00 uH	1.2525 MHz
Rx (spiral)	8325.22	221.00 pF	98.82 uH	1.2494 MHz

Table 1. Summary of measured parameters of each self-resonator.

mutual inductance was obtained as follows Hayes et al. (2003):

$$M_{12} = \frac{|L_{m1} - L_{m2}|}{4}. \qquad (9)$$

Figure 5 shows the theoretical and experimental mutual inductance according to the distance (D) between two self-resonators. In a perpendicular arrangement, the ρ−directed position is fixed at $\rho = 230$ mm. For both coaxial and perpendicular arrangements, the calculated results have good agreement with the measured ones. For the perpendicular case, there is a slight difference between calculation and measurement, especially as the two coils become closer, because the magnetic flux density at each subdivision is not uniform. It can be observed that the mutual inductance for the coaxial case is higher than that for the perpendicular case.

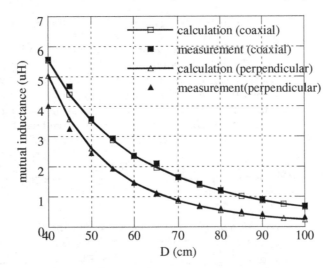

Fig. 5. Calculation and measurement of mutual inductance for both coaxial and perpendicular arrangements.

3.2 Mutual inductance between two rectangular self-resonators
3.2.1 Configuration and derivation
Figure 6 shows a geometrical configuration used to calculate mutual inductance between two rectangular self-resonators arranged in an off-axis position. The Tx coil is placed on the xy plane and its center is $(0,0,0)$. It has N_1 turns. It is L_1 in width and h_1 in height, respectively. The center of the Rx coil is $P_0(x_0, y_0, z_0)$ and the coil is parallel to the y−axis. The Rx coil has N_2 turns. It is L_2 in width and h_2 in height, respectively. Tx and Rx coils are tilted θ degrees. It is assumed that each coil is filamentary. To calculate mutual inductance between Tx and

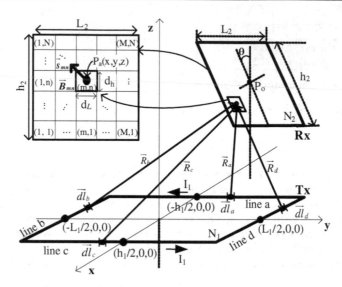

Fig. 6. Configuration of two rectangular self-resonators for calculation of mutual inductance.

Rx self-resonators, the rectangular Tx self-resonator is divided into four lines (line a, line b, line c, line d) and the Rx self-resonator is subdivided (see Figure 6). A subdivision (m, n) is rectangular and its midpoint is $P_a(x, y, z)$. The magnetic flux density at each point $P_a(x, y, z)$ of the subdivision in the Rx self-resonator can be obtained by combining the magnetic flux densities made by the four lines of the Tx self-resonator. It is assumed that $\vec{B}_{mn}$ is uniform in each subdivision. Therefore, by referring to the case of the circular self-resonator in the previous section, the mutual inductance M_{rc} is calculated as follows:

$$M_{12} = M_{rc} = \frac{N_1 N_2}{I_1} \int_{S_2} \vec{B} \cdot d\vec{s}_2 \approx \frac{N_1 N_2}{I_1} \sum_{m=1}^{M} \sum_{n=1}^{N} \vec{B}_{mn} \cdot \vec{s}_{mn}. \qquad (10)$$

$\vec{s}_{mn}$ is the surface of the subdivision (m, n). The magnetic flux density at each subdivision $\vec{B}_{mn}$ is obtained using Bio-Savart's law. The $y-$directed magnetic fields $\hat{y}B_{y_{mn}}$ will not be affected by the mutual inductance due to $\hat{y}B_{y_{mn}} \cdot \vec{s}_{mn} = 0$. Therefore, $\vec{B}_{mn}$ is obtained as follows:

$$
\begin{aligned}
\vec{B}_{mn} &= \frac{\mu_0 I_1}{4\pi} \oint_{C_{Tx}} \left(\frac{d\vec{l'} \times \vec{R}}{R^3} \right) \\
&= \frac{\mu_0 I_1}{4\pi} \left\{ \int_{line\,a} \frac{d\vec{l}_a \times \vec{R}_a}{R_a^3} + \int_{line\,b} \frac{d\vec{l}_b \times \vec{R}_b}{R_b^3} + \int_{line\,c} \frac{d\vec{l}_c \times \vec{R}_c}{R_c^3} + \int_{line\,d} \frac{d\vec{l}_d \times \vec{R}_d}{R_d^3} \right\} \\
&= \hat{x}B_{x_{mn}} + \hat{z}B_{z_{mn}},
\end{aligned}
\qquad (11a)
$$

where

$$
\begin{aligned}
B_{x_{mn}} &= B_x|_{line\,a} + B_x|_{line\,c}, \\
B_{z_{mn}} &= B_z|_{line\,a} + B_z|_{line\,b} + B_z|_{line\,c} + B_z|_{line\,d},
\end{aligned}
\qquad (11b)
$$

with

$$B_x|_{line\,a} = \frac{\mu_0}{4\pi}\left(\frac{-zI_1}{(x+h_1/2)^2+z^2}\right)\left[\left(\frac{y+L_1/2}{R_{a+}}\right)-\left(\frac{y-L_1/2}{R_{a-}}\right)\right],$$

$$B_x|_{line\,c} = \frac{\mu_0}{4\pi}\left(\frac{zI_1}{(x-h_1/2)^2+z^2}\right)\left[\left(\frac{y+L_1/2}{R_{c+}}\right)-\left(\frac{y-L_1/2}{R_{c-}}\right)\right],$$

$$B_z|_{line\,a} = \frac{\mu_0}{4\pi}\left(\frac{I_1(x+h_1/2)}{(x+h_1/2)^2+z^2}\right)\left[\left(\frac{y+L_1/2}{R_{a+}}\right)-\left(\frac{y-L_1/2}{R_{a-}}\right)\right],$$

$$B_z|_{line\,b} = \frac{\mu_0}{4\pi}\left(\frac{I_1(y+L_1/2)}{(y+L_1/2)^2+z^2}\right)\left[\left(\frac{x+h_1/2}{R_{b+}}\right)-\left(\frac{x-h_1/2}{R_{b-}}\right)\right],$$

$$B_z|_{line\,c} = \frac{\mu_0}{4\pi}\left(\frac{-I_1(x-h_1/2)}{(x-h_1/2)^2+z^2}\right)\left[\left(\frac{y+L_1/2}{R_{c+}}\right)-\left(\frac{y-L_1/2}{R_{c-}}\right)\right], \qquad (11c)$$

$$B_z|_{line\,d} = \frac{\mu_0}{4\pi}\left(\frac{-I_1(y-L_1/2)}{(y-L_1/2)^2+z^2}\right)\left[\left(\frac{x+h_1/2}{R_{d+}}\right)-\left(\frac{x-h_1/2}{R_{d-}}\right)\right],$$

$$R_{a+} = R_{b+} = \sqrt{(x+h_1/2)^2+z^2+(y+L_1/2)^2},$$

$$R_{a-} = R_{d+} = \sqrt{(x+h_1/2)^2+z^2+(y-L_1/2)^2},$$

$$R_{c+} = R_{b-} = \sqrt{(x-h_1/2)^2+z^2+(y+L_1/2)^2},$$

$$R_{c-} = R_{d-} = \sqrt{(x-h_1/2)^2+z^2+(y-L_1/2)^2}.$$

Substituting Equation 11a into Equation 10 gives the mutual inductance

$$M_{rc} = \frac{N_1 N_2 d_L d_h}{I_1}\sum_{m=1}^{M=L_2/d_L}\sum_{n=1}^{N=h_2/d_h}\left|-B_{x_{mn}}\cos\theta+B_{z_{mn}}\sin\theta\right|. \qquad (12)$$

3.2.2 Calculation and measurement

To verify the calculation method, four different cases were studied. As shown in Figure 7, the Rx coil was rotated while the Tx coil was fixed. Figure 7a shows that the Rx self-resonator was coaxially arranged with the Tx self-resonator and the center of the Rx self-resonator was $P_0(0,0,D_z)$ while the center for the other cases was $P_0(-20\,\text{cm},0,D_z)$. Figures 7b, 7c, and 7d show that the Tx and Rx self-resonators were tilted $0°$, $90°$, and $45°$, respectively.

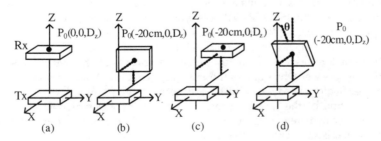

Fig. 7. Schematic drawings of four measurement setups.

In Figure 8, a Tx rectangular self-resonator fabricated in a helical type is illustrated. The Tx and Rx self-resonators were identical. The target resonant frequency was 250 kHz. 14 AWG

Fig. 8. Photograph of a rectangular self-resonator.

litz wire was used for fabrication. The size of the Tx and Rx coils was 62 cm × 33 cm × 5 cm and the number of turns $N_1 = N_2 = 19$. Some results of this study were also presented in Kim et al. (2011).

The intrinsic decay rates, capacitances of high-Q capacitors, measured self-inductances, and measured resonant frequencies of each self-resonator are shown in Table 2.

	Γ	high-Q capacitor	self-inductance	f_0
Tx	1387.5	990.9 pF	407.2 uH	249.85 kHz
Rx	1389.2	984 pF	408.5 uH	250.13 kHz

Table 2. Summary of measured parameters of each rectangular self-resonator.

Figure 9 shows the calculation and measurement results of mutual inductance according to the distance D_z between the Tx and Rx self-resonators. In calculation, the subdivisions were set to be $M = 2000$ and $N = 1000$, that is, $d_L = 0.31$ mm and $d_h = 0.33$ mm.

As shown in Figure 9, it should be pointed out that the calculation had good agreement with the measurement for each case. It is shown that with D_z smaller, the mutual inductance for the coaxial arrangement was higher than the other three cases. It can also be observed that with D_z larger, the mutual inductance for the coaxial arrangement was still the highest, while the mutual inductance for the 0° arrangement was the lowest.

4. Calculation and experimental verification

In order to verify the analysis results and design procedures of an MR-WPT system with an intermediate self-resonator, two MR-WPT systems (coaxial and perpendicular arrangements) were setup as shown in Figure 10. The Tx circular helical self-resonator was the same as that in Figure 4. The Rx self-resonator was identical with the Tx one. A spiral coil as an intermediate circular self-resonator was fabricated to reduce the volume of the MR-WPT system. The measured parameters were the same as those in Table 1. High-Q capacitors were also loaded with each self-resonator in order to adjust the target resonant frequency of each self-resonator and reduce variation of the target resonant frequency by external objects. It should be noted that the intermediate self-resonator was placed at the center between the Tx and Rx self-resonators, that is, the center of the spiral coil was (230 mm, 0, 0). Single loop coils were used as a source coil and a load coil. The transmission coefficient was measured using a vector network analyzer (Agilent 4395A). By varying the distance between the Tx self-resonator and the source coil or the Rx self-resonator and the load coil, the proper impedance matching condition for maximum power transfer efficiency was achieved. It was also found that when

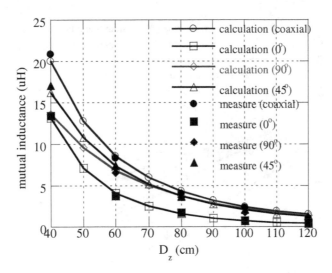

Fig. 9. Calculation and measurement of mutual inductance for both coaxial and perpendicular arrangements.

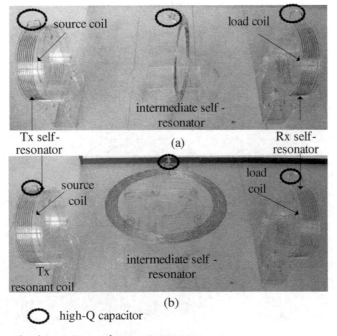

Fig. 10. Photograph of experimental measurement setup.

k_m was nearly five times higher than k_{12} or the distance ($2D_{1m} = D_{1m} + D_{2m}$) was more than 80 cm, k_{12} can be negligible (see Kim et al. (2011)).

In Figure 11, the measured and calculated efficiencies of two MR-WPT systems with coaxial

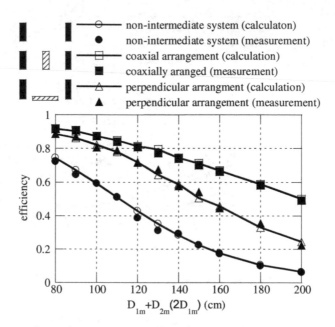

Fig. 11. Measured and calculated efficiencies of MR-WPT systems without an intermediate self-resonator and with coaxially arranged and perpendicularly arranged intermediate self-resonators vs. distance.

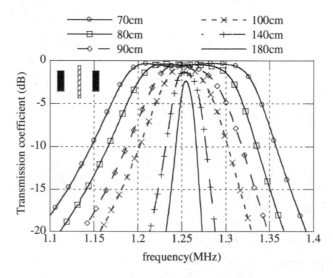

Fig. 12. Efficiency measurement of the MR-WPT system with the coaxially arranged intermediate self-resonator vs. frequency.

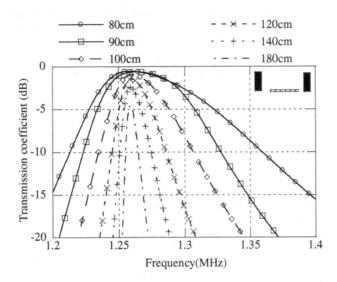

Fig. 13. Efficiency measurement of the MR-WPT system with the perpendicularly arranged intermediate self-resonator vs. frequency.

or perpendicular arrangements according to the distance between Tx and Rx self-resonators are displayed. In addition, the efficiency of an MR-WPT system without the intermediate self-resonator is displayed to make a comparison with the systems with the intermediate coil. With the aid of Equation 4, the efficiencies of the systems were calculated, and the measured parameters in Table 1 were used for calculation. As shown in Figure 11, the experimental and theoretical results were very consistent. The efficiency for the coaxial arrangement case was higher than that for the perpendicular arrangement, because the mutual inductance of the coaxial arrangement was higher as shown in Figure 5. It should be noted that the system with the intermediate self-resonator has higher efficiency than that without the intermediate self-resonator. This means that using intermediate self-resonators with low losses can help to improve power transfer efficiency and extend the coverage of wireless power transfer effectively.

Figures 12 and 13 show the measured efficiencies of the coaxial and perpendicular arrangement systems for different distances according to frequency, respectively. In the case of the coaxial arrangement, the efficiencies were nearly symmetric according to frequency while those for the case of the perpendicular arrangement were asymmetric according to frequency. The reason for this was that with a shorter distance in the perpendicular arrangement case, k_m was no higher than k_{12} and k_{12} was no longer negligible. It should also be pointed out that using intermediate self-resonators can help to make the operating frequency bandwidth broader.

5. Conclusion

In this article, the characteristics of an MR-WPT system with intermediate self-resonators were analyzed. Its power transfer efficiency was derived and the matching condition for maximum power transfer was also obtained. The calculation methods of mutual inductance between two circular or rectangular self-resonators were presented and some calculation results were also explained. The analysis results, calculation methods, and design procedures were verified by experimental measurement. The measurements and calculations show that if intermediate self-resonators are properly used, an MR-WPT system with intermediate self-resonators transfers wireless power efficiently up to several meters. In particular, it is shown that the efficiency of an MR-WPT system with two self-resonators arranged perpendicularly is as good as that of a coaxially arranged MR-WPT system within a certain area. Therefore, it is expected that these analysis results and properties of an MR-WPT system with intermediate self-resonators can be well applied to develop various applications.

6. References

Kurs, A.; Karalis, A.; Moffatt, R.; Joannopoulos, J. D.; Fisher, P. & Soljačić, M. (2007). Wireless power transfer via strongly coupled magnetic resonances, *Science*, Vol.317, July 2007, pp. 83–86, ISSN 0036-8075.

Karalis, A.; Joannopoulos, J. D. & Soljačić, M. (2008). Efficient wireless non-radiative mid-range energy transfer, *Annals of Physics*, 323, 2008, pp. 34–48, ISSN 0003-4916.

Sample, A. P.; Meyer, A. & Smith, J. R. (2010). Analysis, experimental results, and range adaption of magnetically coupled resonators for wireless power transfer, *IEEE Transactions on Industrial Electronics*, Vol.58, No. 2, 2010, pp. 544–554, ISSN 0036-8075.

Hamam, R. E.; Karalis, A.; Joannopoulos, J. D. & Soljačić, M. (2009). Efficient weakly-radiative wireless energy transfer:an EIT-like approach, *Annals of Physics*, 324, 2009, pp. 1783–1795, ISSN 0003-4916.

Haus, H. A. (1984). *Waves and fields in optoelectronics*, Prentice Hall, pp. 197–234, ISBN 0-13-946053-5, NJ.

Good, R. H. (2001). Elliptic integrals, the forgotten functions, *European Journal of Physics*, Vol.22, 2001, pp. 119–126, ISSN 0142-0807.

Babic, S.; Sirois, F.; Akyel, C. & Girardi, C. (2010). Mutual inductnace calculation between circular filaments arbitrarily positioned in space: alternative to Grover's formula, *IEEE Transactions on Magnetics*, Vol.46, No. 9, 2010, pp. 3591–3600, ISSN 0018-9464.

Hayes, J. G.; O'Donovan, N.; Egan, M. G. & O'Donnell, T. (2003). Inductance characterization of high-leakage transformers, *IEEE Applied Power Electronics Conference and Exposition (APEC)*, pp. 1150–1156, ISBN 0-7803-7768-0, Feb. 2003, IEEE, FL, USA.

Kim, J. W.; Son, H. C.; Kim, D. H.; Kim, K. H. & Park, Y. J. (2011). Efficiency of magnetic resonance WPT with two off-axis self-resonators, *IEEE MTT-s international microwave workshop series on innovative wireless power transmission (IMWS-IWPT 2011)*, pp. 127–130, ISBN 978-1-61284-215-8, May 2011, IEEE, Kyoto, Japan.

Kim, J. W.; Son, H. C.; Kim, K. H. & Park, Y. J. (2011). Efficiency analysis of magnetic resonance wireless power transfer with intermediate resonant coil, *IEEE Antennas and Wireless Propagation Letters*, Vol.10, 2011, pp. 389–392, ISSN 1536-1225.

5

Equivalent Circuit and Calculation of Its Parameters of Magnetic-Coupled-Resonant Wireless Power Transfer

Hiroshi Hirayama
Nagoya Institute of Technology
Japan

1. Introduction

Because of an improvement of wireless communication technologies, cables are taken away step by step from an electrical equipment. Now, a power cable is the last wire connected to the equipment. Therefore, wireless power transfer (WPT) technology is desired.

Conventionally, microwave power transfer and magnetic induction have been used for this purpose. Microwave power transfer technology utilizes a directivity antenna with high-power microwave generator. This technology is applicable for very long range power transfer, such as space solar power system(SPS). However, this technology requires precise directivity control. Thus it is difficult to use in consumer electrical appliances.

Magnetic induction technology is already used for power supply for a cordless phone, electric tooth blush, or electric shaver. RFID tag also utilizes magnetic induction technology for power supply. However, demerit of this technology is very short distance for power transmission. Although transmitter and receiver is electrically isolated, they should be physically touched.

Wireless power transfer using magnetic-coupled resonance is proposed by MIT in 2007 (1). 60W power transmission experiment for 2m distance was demonstrated (2). It is amazing that this technology is applicable although human body is located just between a transmitter and receiver. Thus, it is expected that this technology is used in home electrical appliances or electric vehicles. On the other hands, for practical implementation, design criteria of this system should be established.

Although the magnetic-coupled resonant WPT technology is the epoch-making technology, this technology seems similar technology to magnetic induction. Both of them use magnetic field to transfer power and magnetic coil to feed/pickup power. The key point of the magnetic-coupled resonant WPT is using capacitance to occur resonance. However, even in conventional magnetic induction technology, capacitor is also used. The feature of the magnetic-coupled resonant WPT is to utilize odd-mode and even-mode resonance. In the magnetic induction technology, frequency is fixed according to the transfer distance. However in the magnetic-coupled WPT technology, frequency should be varied according to the transfer distance.

In the section 2, we explain basic theory of magnetic-coupled WPT system through equivalent circuit expression. Method of moment (MoM) simulations will be demonstrated for validation. In the section 3, a procedure to calculate parameters of the equivalent circuit of magnetic coupled resonant WPT is discussed. Equivalent circuit parameters are calculated from the geometrical and material parameters of the WPT structure. Radiation loss and conductive loss are taken into account in the equivalent circuit. Both direct-fed type and indirect-fed type structures are considered. MoM calculation shows that the equivalent circuit has capability to calculate S parameters and far-field radiation power correctly.

2. Basic theory of magnetic-coupled resonant WPT system

2.1 Classification of coupled resonant WPT system

The coupled resonant WPT system is classified from the viewpoint of 1) field used in coupling, 2) power feeding, and 3) resonant scheme.

From the viewpoint of field, this system is classified to magnetic-coupled resonant type and electric-coupled resonant type. Not only magnetic field but also electric field can be used for coupled-resonant WPT. In the case of magnetic-field coupling, power transfer is affected by surrounding permeability. On the other hand, for electric-field coupling, power transfer is affected by surrounding permittivity. Because the relative permeability of human body is almost unity, magnetic-coupled resonant WPT is not affected by human body. Therefore, magnetic-coupled type is used in the reference (2). However, when the effect of a human body is negligible, electric-field coupling type may be a considerable alternative.

From the viewpoint of power feeding, this system is classified to direct fed type and indirect fed type. In the case of the direct fed type, power source and load is directly connected to the resonant structure. On the other hand, feeding loop is used for indirect fed type. A power source and load is connected to the loop structures, which have not sharp frequency characteristics. The loop structure and the resonant structure is coupled by magnetic induction. A merit of direct fed type is simple structure. On the other hand, indirect type has advantage in freedom of design: impedance matching is achieved by adjusting the spacing between the loop and the resonant structure.

From the viewpoint of resonant scheme, this system is classified to self resonant type and external resonant type. To occur resonance, same quantity of inductive reactance and capacitive reactance is necessary. For the self resonant type, inductance and capacitance are realized by identical structure. For example, open-end spiral structure is used. Currents along the spiral structure causes inductance and charges at the end of the spiral causes capacitance. Resonant occurs at the frequency at which the inductive and the capacitive reactance becomes same. On the other hands, external resonant type has distinct structure to realize capacitance and to realize inductance. For example, loop structure is used for inductance and discrete capacitor is added for capacitance.

In accordance with the above discussed classification, the WPT system shown by MIT (2) is magnetic-coupled resonant, indirect-fed, self resonant type.

Although this system is regarded to be a new technology, its principle can be discussed from following standpoints. 1) this system is a TX antenna and a RX antenna. 2) this system is two element TX array antenna which has parasitic element terminated with a resistor. 3) this system is coupled resonators. 4) this system is a transformer with capacitors. The viewpoint 1) is useful to discuss impedance matching or power transfer efficiency. However, since this system is used in near-field region, a concept based on far-field region, such as gain or directivity, is not applicable. The viewpoint 2) is beneficial to consider far-field emission or interaction between TX and RX antenna. From a viewpoint of electro-magnetic

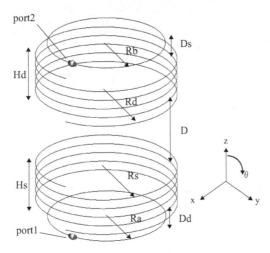

Fig. 1. Calculation model

	Ds	D	Dd	Hs	Hd	Ra	Rb	Rs	Rd
[mm]	50	300	50	200	200	250	250	300	300

Table 1. Parameters of the calculation model

resonant mode, the viewpoint 3) provides considerable knowledge, as in the reference (2). The viewpoint 4) is profitable to predict input impedance or resonant frequency from an equivalent circuit. On the other hands, this is unsuitable to discuss undesired emission or effect to human bodies from a viewpoint of electromagnetics. In this section, we unveil mechanism of power transfer from a suitable standpoint.

2.2 Investigation on resonant mechanism
A consideration model is shown in Fig. 1. Structural parameters are listed in Table 1. This model is used in the reference (2), which is classified magnetic-coupled resonant, indirect-fed, self resonant type. The loop structure is used to feed. The helical structure is used to occur resonance. The loop structure and the helical structure is coupled by magnetic inductance. The helices are coupled by magnetic-coupled resonance.

This structure is analyzed by MoM calculation. Perfect electric conductor (PEC) is assumed in this calculation. Figure 2(a) shows input impedance. The case without helix is also plotted. By using the helix, two resonances are occurred.

Figure 2(b) shows real part and imaginary part of the port current. Port 1 is input port and Port 2 is output port. We can see that at the lower frequency resonance, input port and output port have same polarity of current. On the other hand, at the higher frequency resonance, input port and output port have opposite polarity of current.

Magnetic field distributions at these frequencies are shown in Fig. 3 and 4. Sub-figure (a) shows magnitude of magnetic field vector, (b) shows z-direction component of magnetic field vector, which corresponds to vertical direction in the figure, and (c) shows y-direction component, which corresponds to horizontal direction in the figure. At the lower frequency

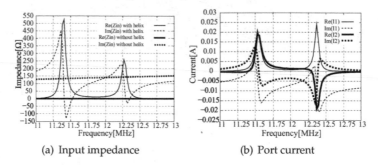

(a) Input impedance   (b) Port current

Fig. 2. Frequency characteristics of the consideration model

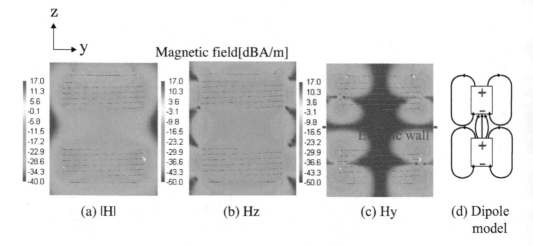

(a) |H| (b) Hz   (c) Hy   (d) Dipole model

Fig. 3. Magnetic field distribution on a frequency of the odd mode(11.4MHz) and equivalent dipole model

resonance, there is no y component magnetic field at the center of TX and RX. Therefore electric wall is located between TX and RX. Considering the polarity of port current, this resonant mode is modeled by magnetic dipole shown in sub-figure(d). Since the magnetic charge distribution along z axis is asymmetrical, this resonant mode can be called odd mode. On the other hand, at the higher frequency resonance, there is no z component magnetic field at the center of TX and RX. Therefore magnetic wall is located between TX and RX. Since the magnetic charge for this frequency is symmetrical, this resonant mode can be called even mode.

Next, let us see power transmission efficiency. Power transmission efficiency is defined by

$$\eta = \frac{Z_l \, |I_2|^2}{|I_1||V_1|}. \tag{1}$$

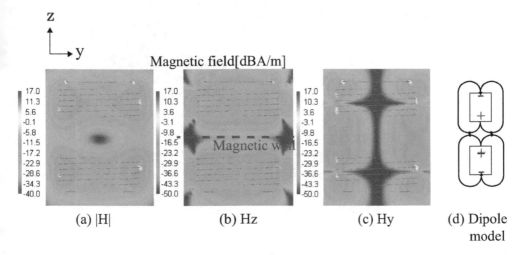

(a) |H| (b) Hz   (c) Hy   (d) Dipole
                                                                     model

Fig. 4. Magnetic field distribution on a frequency of the even mode(12.3MHz) and equivalent dipole mode

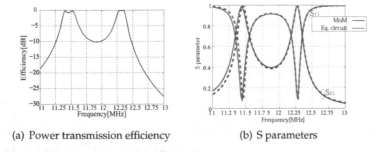

(a) Power transmission efficiency   (b) S parameters

Fig. 5. Frequency characteristics of the transmission specifications

Figure 5(a) shows the calculated result. There are two peaks of efficiency both for odd and even mode. The imaginary part of the input impedance (Fig. 2(a)) is not zero at the frequency of resonance since the input impedance includes the inductance of the loop structure. Therefore, the resonant frequency and the frequency at which the input impedance becomes zero become different. From the viewpoint of power engineering, this definition means effective power of the output port normalized by apparent power of the input port. It is said that the power transmission efficiency is maximized when the power factor becomes unity. From the viewpoint of radio frequency engineering, S parameter is important index. Figure 5(b) shows calculated S parameters. At the odd and even mode frequencies, S_{21} becomes almost unity.

We assume a sphere which surrounds the transmitting and receiving antenna. By applying the Poynting theorem to the volume of the sphere V and the surface of the sphere S, a balance of the power is described as

$$P_p = P_r + P_w + P_d \tag{2}$$

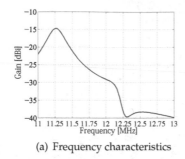

(a) Frequency characteristics

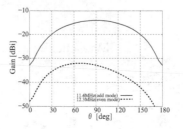

(b) Angle characteristics

Fig. 6. Gain as transmitting antenna

where

$$P_p = \iiint_V P_s dV \tag{3}$$

$$= P_{in} - P_{out} \tag{4}$$

$$P_r = \iint_S \mathbf{S} \cdot \mathbf{n} dS \tag{5}$$

$$P_w = \frac{\partial}{\partial t} \iiint_V \left(\frac{1}{2}\mu \mathbf{H}^2 + \frac{1}{2}\varepsilon \mathbf{E}^2 \right) dV \tag{6}$$

$$P_d = \iiint_V \sigma \mathbf{E}^2 dV \tag{7}$$

and P_{in} is a power supplied to the port of the transmitting antenna, P_{out} is a power extracted from the port of the receiving antenna, P_r is a power of far-field emission, P_w is stored energy in the volume V, and P_d is loss power dissipated in the volume V. For an antenna used for far-field, an antenna is a device which transduces P_p into P_r. However, since the wireless power transmission is a system which extracts P_{out} from the P_{in}, P_r becomes a loss energy.

An antenna used in far-field is required to maximize far-field emission, i.e. to maximize a gain. In the wireless power transmission, far-field emission becomes not only power loss but also a cause of electro-magnetic interference (EMI). Thus, reducing far-field emission is required. The far-field emission can be estimated from the gain by regarding this system as an array antenna consisted of the TX antenna and the RX antenna. In this calculation, RX antenna is considered as a parasitic element loaded with a loss resistance located in the vicinity of the TX antenna. Fig. 6(a) shows the gain as antenna for x direction. Fig. 6(b) shows angle specifications of gain for the odd and the even mode frequencies. The maximum gain was -13.8 dBi and -31.8 dBi for odd and even mode, respectively. The gain of the even mode frequency for the maximum emission direction is smaller by 18 dB than that of the odd mode frequency.

Strength of undesired emission is specified by using effective isotropic radiation power (EIRP):

$$\text{EIRP} = G_t P_t \tag{8}$$

where G_t and P_t show gain of antenna shows power fed to the antenna, respectively. EIRP for the 1W power transfer is 41.7mW for odd mode and 0.66mW for even mode. Since the current of the TX antenna and RX antenna has opposite direction for the even mode, the resultant

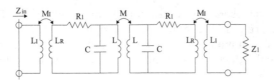

Fig. 7. Equivalent circuit model

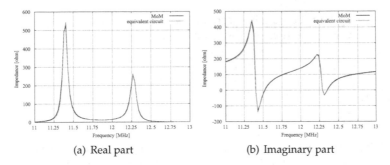

(a) Real part   (b) Imaginary part

Fig. 8. Input impedance calculated by MoM and the equivalent circuit model

emission from the TX and RX antennas is canceled each other in far field. This result shows that the even mode is suitable for use because of low undesired radiation.

2.3 Investigation on resonant frequency with equivalent circuit model

Equivalent circuit of the wireless power transmission is shown in Fig. 7. L, C, and M represent self inductance of the helix structure, stray capacitance of the helix structure, and the mutual inductance between transmitting and receiving helices, respectively R_l and Z_l represent loss resistance of the helix structure and load impedance, respectively. L_r and L_I correspond to self inductance of the induction coil and self inductance of the helix which concerns with the coupling to the induction coil, respectively. M_I shows the mutual inductance between induction coil and resonant helix. Figure 8(a) and 8(b) show input impedance calculated by the MoM and the equivalent circuit, respectively. We can see that the physical phenomenon of WPT is adequately represented by the equivalent circuit.

Because the loop structure does not concern with a resonant mechanism, resonant frequency determined by the L, M, and C will be considered. Resonant frequency of the helical structure f_0 becomes

$$f_0 = \frac{1}{2\pi\sqrt{LC}}. \tag{9}$$

When two helixes are coupled each other, odd mode resonance and even mode resonance occur. Resonant frequency of odd mode and even mode, f_{odd} and f_{even}, becomes as follows:

$$f_{odd} = \frac{1}{2\pi\sqrt{C(L+M)}} \tag{10}$$

$$f_{even} = \frac{1}{2\pi\sqrt{C(L-M)}}. \tag{11}$$

In the case of odd mode resonant, because polarity of current of the TX and the RX coil is same, mutual inductance is added to the mutual inductance. Thus, f_{rmodd} becomes smaller than f_0. In the case of even mode resonant, because polarity of current of the TX and the RX coil is opposite, mutual inductance is substituted from the mutual inductance. Thus, f_{rmodd} becomes higher than f_0.

Coupling coefficient k is defined by

$$k = \frac{M}{L},\tag{12}$$

which shows strength of coupling between the TX and the RX loop. By using the k, the odd and the even mode resonant frequency is obtained by

$$f_{odd} = \frac{f_0}{\sqrt{1+k}}\tag{13}$$

$$f_{even} = \frac{f_0}{\sqrt{1-k}}.\tag{14}$$

For the traditional WPT system using magnetic induction, frequency is fixed with respect to the distance between TX and RX. However, for the magnetic-coupled resonant WPT system, resonant frequency varies with respect to the distance between TX and RX. In the practical use, available frequency is determined due to regulation. To maintain resonant frequency, capacitance should be adaptively changed.

f_0 and k is obtained when the resonant frequency of the odd and the even mode is known:

$$f_0 = \sqrt{\frac{2f_{even}^2 f_{odd}^2}{f_{even}^2 + f_{odd}^2}}\tag{15}$$

$$k = \frac{f_{even}^2 - f_{odd}^2}{f_{even}^2 + f_{odd}^2}.\tag{16}$$

To estimate f_0 and k from an experimental result, this equations can be used.

3. Calculation of equivalent circuit parameters

3.1 Consideration model

Fig. 9 shows a structure of the direct-fed type wireless power transfer. Both transmitting (TX) and receiving (RX) antennas consist of one-turn loop whose radius is r. The circumference of the loop is assumed to be much smaller than the wavelength of the operating frequency. The loops are made of wire whose diameter is $2d$ and conductivity is σ. Distance between TX and RX is h. Capacitor C_0 is connected to occur resonance.

Equivalent circuit of this structure is shown in Fig. 10. Z_S and Z_L show source impedance and load impedance, respectively. L and M show self and mutual inductance of the loops, respectively. R_r and R_l show radiation resistance and conductive loss resistance, respectively. Fig .11 shows a consideration model of the indirect-fed type. Loops 1 and 2 are for TX, and 3 and 4 are for RX. Loops 1 and 4 are feeding coils, which are connected to the power source and the load, respectively. Loops 2 and 3 are resonant coils, to which the resonant capacitors

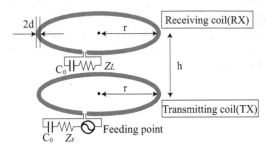

Fig. 9. Consideration model of direct-fed type WPT structure

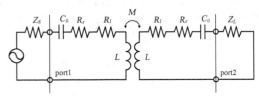

Fig. 10. Equivalent circuit of the direct-fed model

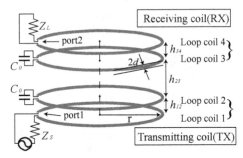

Fig. 11. Consideration model of indirect-fed type WPT structure

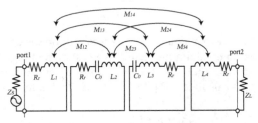

Fig. 12. Equivalent circuit of the indirect-fed model

are connected. Power is transferred by magnetic induction between the feeding coil and the resonant coil. Between resonant loops, power is transferred by magnetic coupled resonant. This WPT system is characterized by the geometrical dimensions: loop radius r, radius of wire d, distance between loops h_{12}, h_{23}, and h_{34}.

Equivalent circuit of this structure is shown in Fig. 12. L_i shows self inductance of i-th loop. M_{ij} shows mutual inductance between i-th and j-th loops. In this consideration, all combination of mutual coupling is considered. Equivalent circuit parameters are calculated in the same manner as the direct-fed type.

3.2 Equivalent circuit parameters

By using the Neumann's formula, mutual inductance M becomes

$$M = \frac{\mu_0}{4\pi} \oint_{1} \oint_{2} \frac{\vec{dl_1} \cdot \vec{dl_2}}{r_{12}} \tag{17}$$

where μ_0 shows permeability of free space, dl_1 and dl_2 show line element of TX and RX loops, respectively, and r_{12} shows distance between dl_1 and dl_2. Assuming that two loops are arranged as shown in Fig. 9, M is reduced to

$$M = \mu_0 \sqrt{r_1 r_2} \left\{ \left(\frac{2}{k} - k \right) K(k) - \frac{2}{k} E(k) \right\} \tag{18}$$

$$\tag{19}$$

where

$$k = \frac{4 r_1 r_2}{(r_1 + r_2)^2 + h^2}$$

and $K(k)$ and $E(k)$ are complete elliptic integrals:

$$K(k) = \int_0^{\pi/2} \frac{1}{\sqrt{1 - k^2 \sin^2 \phi}} d\phi \tag{20}$$

$$E(k) = \int_0^{\pi/2} \sqrt{1 - k^2 \sin^2 \phi} d\phi. \tag{21}$$

Self inductance L is calculated from external inductance L_e and internal inductance L_i:

$$L = L_e + L_i \tag{22}$$

The external inductance is due to magnetic field caused around the conductor. Assuming that all the current I is concentrated along the center line C shown in Fig. 13, resultant magnetic flux inside the C is identical to that inside the C'. Thus, by substituting $r_1 = r$, $r_2 = r - d$, $h = 0$ into the Eq. (18), L_e is calculated. Eq. (18) is approximated to

$$L_e \approx \mu_0 r \left(\ln \frac{8r}{d} - 2 \right) \tag{23}$$

assuming $r_1 \approx r_2$, $h \ll r_1, r_2$.

The internal inductance is caused by magnetic fields in the conductor. If the loop is perfect conductor, currents flow only on the surface of the conductor, then internal inductance is

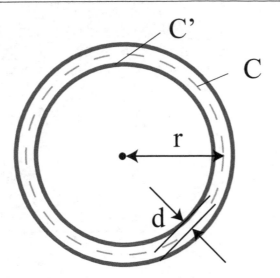

Fig. 13. External inductance L_e

negligible. Considering the conductivity, skin depth δ is calculated from

$$\delta = \frac{1}{\sqrt{\omega \mu \sigma}}. \tag{24}$$

Internal inductance becomes

$$L_i = \frac{\Phi_i}{I} = 2\pi r \frac{\int_{d-\delta}^{d} B_i dr'}{I} \tag{25}$$

where I, Φ_i, B_i and μ show current in the conductor, flux linkage, flux density in the conductor, permeability of the conductor, respectively. B_i is obtained from Ampere's law:

$$B_i = \mu \frac{NI}{2\pi r'} \tag{26}$$

where $r'(d - \delta < r' < d)$ shows distance from the center of the conductor. Current density in the conductor becomes uniform assuming $\delta > d$, thus internal inductance becomes

$$L_i = \frac{\mu r}{4}. \tag{27}$$

Because the circumference is much smaller than the wavelength, radiation resistance R_r is approximated to

$$R_r = 20\pi^2 (\beta r)^4 \tag{28}$$

where $\beta = 2\pi/\lambda$.

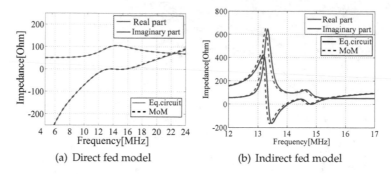

(a) Direct fed model   (b) Indirect fed model

Fig. 14. Input impedance

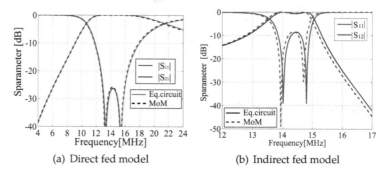

(a) Direct fed model   (b) Indirect fed model

Fig. 15. S parameters

Taking the skin effect into account, conductive loss is obtained by

$$R_l = \frac{2\pi r}{\sigma \cdot S} \tag{29}$$

where $S = \pi\{d^2 - (d - \delta)^2\}$ is cross section in which current flows.
Finally, resonant capacitance becomes

$$C_0 = \frac{1}{(2\pi f_0)^2 L}. \tag{30}$$

3.3 Numerical validation
To discuss adequateness of the equivalent circuit, input impedance and S parameters calculated by the proposed procedure are compared with those calculated by MoM. Geometrical dimensions are set to be $r = 20$cm, $d = 1$mm, and $h = 2$cm. Resonant frequency is designed to be 13.2MHz. Figure 14(a) shows the real part and the imaginary part of the input impedance. Figure 15(a) shows frequency characteristics of S parameters. From the S_{11}, two resonant modes, that is odd mode and even mode, are confirmed. It is proved that the procedure to calculate equivalent circuit parameters is adequate.

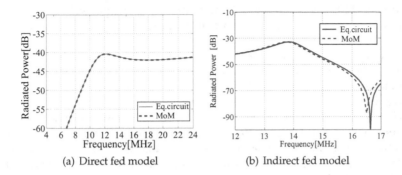

(a) Direct fed model   (b) Indirect fed model

Fig. 16. Far field radiation power

In the case of the indirect-fed model, input impedance and S parameters are calculated through impedance matrix $\mathbf{Z}$:

$$\mathbf{Z} = j\omega\mathbf{Z}_I + \mathbf{Z}_E + \mathbf{Z}_R \tag{31}$$

where

$$\mathbf{Z}_I = \begin{bmatrix} L_1 & M_{12} & M_{13} & M_{14} \\ M_{12} & L_2 & M_{23} & M_{24} \\ M_{13} & M_{23} & L_3 & M_{34} \\ M_{14} & M_{24} & M_{34} & L_4 \end{bmatrix} \tag{32}$$

$$\mathbf{Z}_E = \text{diag}\left(Z_S, \frac{1}{j\omega C_0}, \frac{1}{j\omega C_0}, Z_L \right) \tag{33}$$

$$\mathbf{Z}_R = (R_L + R_r)\mathbf{I}. \tag{34}$$

and $\text{diag}(\cdot)$ and $\mathbf{I}$ show diagonal matrix and identity matrix, respectively.
Far-field radiation power is calculated from

$$P_r = \frac{1}{2}\left(R_r|I_1 + I_2 + I_3 + I_4|^2 \right). \tag{35}$$

Figure 14(b) and 15(b) shows calculation result. MoM result almost agreed with the result of the equivalent circuit. However, in comparison with the direct-fed type, error can be seen. In the case of the indirect-fed type, the feeding loop is closed to the resonant loop. It is considered that this error is caused by stray capacitance between feeding loop and resonant loop.
In the equivalent circuit, total power dissipated in the radiation resistance becomes

$$P_r = \frac{1}{2}\left(R_r|I_1|^2 + R_r|I_2|^2 \right). \tag{36}$$

However, far-field radiation due to TX and RX loops is added in field strength, not in power. Thus, radiation power P_r is calculated considering phase of current:

$$P_r = \frac{1}{2}\left(R_r|I_1 + I_2|^2 \right). \tag{37}$$

Figure 16(a) and 16(b) shows calculated result. In this graph, power is normalized by the incident power to the port 1. It is shown that the proposed procedure has a capability to estimate far-field radiation power correctly.

Figure 17 shows dependency of S parameters on distance of transmitting at the resonant frequency. It is confirmed that the equivalent circuit represent S parameters correctly.

Figure 18 shows effect of conductivity on S parameters. It is shown that conductivity of material is adequately taken account in the equivalent circuit.

To discuss the limitation of the equivalent circuit, we calculated S parameters with respect to the radius of the loop. Figure 19 shows S parameter of radius 10cm($2\pi r$=0.0276λ), 30cm($2\pi r$=0.0830λ), 40cm($2\pi r$=0.1106λ), 50cm($2\pi r$=0.1382λ). When the radius of the loop becomes larger, the MoM result is not in accordance with the result of the equivalent circuit. It is considered that when the loop becomes large, circumference is not negligible compared to the wavelength then current distribution on the coil becomes not uniform.

3.4 Experimental validation

To validate simulation result, experimental model was fabricated. Figure 20(a) shows configuration of the experimental model. Copper wire and 100pF ceramic capacitors were used.

Measured S parameters using VNA are shown in Fig.20(b). S parameters obtained by the equivalent circuit are also plotted. For the equivalent circuit, conductivity is fit to consistent with the experimental value. Conductivity of 1.90×10^4[1/m $\cdot\Omega$] gave most suitable S

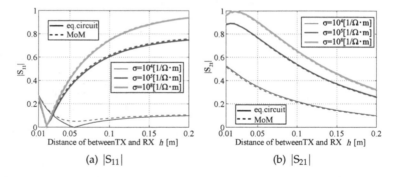

(a) $|S_{11}|$ (b) $|S_{21}|$

Fig. 17. Effect of transfer distance on the S parameters

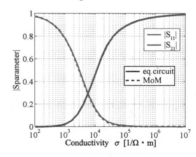

Fig. 18. Effect of the conductivity on the S parameters

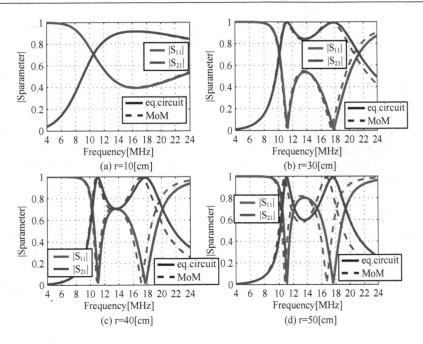

Fig. 19. Frequency characteristics of the S parameters($\sigma = 10^8[1/\text{m} \cdot \Omega]$, $h = r/16$)

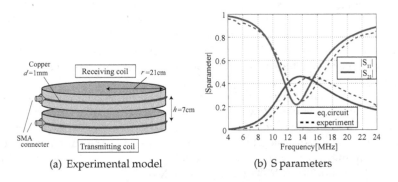

(a) Experimental model   (b) S parameters

Fig. 20. Experimental validation

parameters for the experimental result whereas literature value is $6.80 \times 10^7 [1/\text{m} \cdot \Omega]$. It is considered that a cause of the error is dielectric loss of the capacitor.

4. Conclusion

In this chapter, resonant mechanism of magnetic-coupled resonant wireless power transfer was demonstrated. There are odd mode and resonant mode resonance. In the odd mode resonance, port current of input port and output port are opposite polarity. Electric wall is formed between TX and RX. Resonant frequency becomes lower because total inductance becomes self inductance plus mutual inductance. Since far-field radiation from TX and RX are in-phase, undesired emission becomes large. In the even mode resonance, port current of input port and output port are same polarity. Magnetic wall is formed between TX and RX. Resonant frequency becomes higher because total inductance becomes self inductance minus mutual inductance. Since far-field radiation from TX and RX are out-of-phase, undesired emission becomes small. From the viewpoint of low undesired emission, even mode resonant is better to use.

Next, equivalent circuit of WPT system and procedure to calculate its parameters from geometrical dimensions was shown. Both radiation loss and conductive loss are taken into account in the equivalent circuit. MoM calculation shows that the equivalent circuit has capability to calculate S parameters and far-field radiation power correctly. It is expected to utilize the equivalent circuit to design a matching network.

5. References

[1] Andre Kurs, Arsteidis Karalis, Robert Moffatt, John joannpoulos, Peter Fisher, Marin Soljacic, "Wireless Power Transfer via Strongly Coupled Magnetic Resonances," Science Magazine, Vol.317, No.5834, pp.83-86, 2007.

[2] Arsteidis Karalis, J. D. joannpoulos, Marin Soljacic, "Efficient wireless non-radiative mid-range energy transfer," Annals of Physics, Vol. 323. No. 1, pp. 34-48, Jan. 2008.

[3] Hiroshi Hirayama, Toshiyuki Ozawa, Yosuke Hiraiwa, Nobuyoshi Kikuma, Kunio Sakakibara, "A Consideration of Electromagnetic- resonant Coupling Mode in Wireless Power Transmission,"IEICE ELEX, Vol. 6, No. 19, pp. 1421-1425, Oct. 2009

[4] Hiroshi Hirayama, Nobuyoshi Kikuma, Kunio Sakakibara, "An consideration on equivalent circuit of wireless power transmission,"Proc. of Antem, Jul. 2010

[5] Hiroshi Hirayama, Nobuyoshi Kikuma, Kunio Sakakibara, "Undesired emission from Magnetic-resonant wireless power transfer," Proc. of EMC Europe, Sep. 2010

[6] I. Awai, "Design Theory of Wireless Power Transfer System Based on Magnetically Coupled Resonators," Proc. 2010 IEEE International Conference on Wireless Information Technology and Systems, Aug. 2010.

[7] Qiaowei Yuan, Qiang Chen, Kunio Sawaya, "Transmission Efficiency of Evanescent Resonant Coupling Wireless Power Transfer System with Consideration of Human Body Effect," IEICE Tech. Report, vol. 108, no. 201, AP2008-91, pp. 95-100, Sep. 2008.

[8] Koichi Tsunekawa, "A Feasibility Study of Wireless Power Transmission using Antenna Mutual Coupling Technique on Indoor Ubiquitous Wireless Access System," IEICE Tech. Report, vol. 108, no. 304, AP2008-113, pp. 13-18, Nov. 2008.

[9] Masato Tanaka, Naoki Inagaki, Katsuyuki Fujii, "A new wireless connection system through inductive magnetic field," IEICE Tech. Report, vol. 108, no. 386, AP2008-184, pp. 197-202, Jan. 2009.

6

Performance Analysis of Magnetic Resonant System Based on Electrical Circuit Theory

Hisayoshi Sugiyama
Dept. of Physical Electronics and Informatics
Osaka City University
Japan

1. Introduction

In this chapter, performances of wireless resonant energy links(Karalis et al., 2008) based on nonradiative magnetic field(Kurs et al., 2007) generated by a pair of coils are analyzed using electrical circuit theory. (from now on, this type of energy link is simply called as *magnetic resonant system*.) Based on the analyses, several aspects of the magnetic resonant systems concerning their power transmission characteristics become clear. In addition, optimal design parameters of the system are obtained that maximize the power transmission efficiency or effective power supply.

Simple and common processes based on the electrical circuit theory, other than specialized ones such that coupled-mode theory(Haus, 1984) or antenna theory(Stutzman, 2011), are applied to the analyses throughout this chapter. Therefore, intuitive and integrated comprehension of the magnetic resonant systems may be possible especially in the comparison of principle with the conventional electromagnetic induction systems as described in the next section.

In section 2, the difference of principle of magnetic resonant systems from that of conventional electromagnetic induction systems is explained where the role of magnetic resonance become clear when the coils are located apart. In section 3, based on the difference from the conventional systems, the inherent characteristics of power transmission of magnetic resonant systems are explained focusing on the resonance current in coils that affects the efficiency of the power transmission. In section 4, based on the power transmission characteristics of magnetic resonant systems, optimal system designs are investigated that maximize the power transmission efficiency or effective power supply under constraints of coil sizes and voltages of power sources. Finally, section 5 summarizes these discussions.

2. Electromagnetic induction system and magnetic resonant system

In this section, first, the principle of conventional electromagnetic induction systems is explained based on the electrical circuit theory. In comparison with this principle of the conventional systems, second, that of magnetic resonant systems is explained where the role of magnetic resonance become clear in power transmission between distant coils.

2.1 Electromagnetic induction system

An electromagnetic induction system is commonly used as a voltage converter that consists of closely coupled pair of coils with different numbers of wire turns. In this system, alternating voltage applied to one part of coil induces a different terminal voltage of the counterpart at the same frequency. Because mutual inductance of coils is so large by the proximity of them even with iron core that constrains the magnetic flux within their inner field, the adequate high power transmission efficiency is obtained by this conventional system.

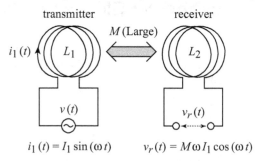

$$i_1(t) = I_1 \sin(\omega t) \qquad v_r(t) = M\omega I_1 \cos(\omega t)$$

Fig. 1. Electromagnetic induction system.

Figure 1 shows a simple model of the electromagnetic induction system consisting of a pair of coils with different inductances L_1 and L_2. They are closely coupled and have large mutual inductance M. No load resistance is connected to the receiver coil. In this system, the transmitter coil is driven by a voltage source $v(t)$ and current $i_1(t) = I_1 \sin(\omega t)$ flows through its wire at some stationary state of the system. Almost of the alternating magnetic flux generated by this current interlinks the receiver coil. This flux linkage Φ is derived by $i_1(t)$ and the mutual inductance M of coils as $\Phi = Mi_1(t)$. Because the differential of the flux linkage Φ equals the induced voltage of the coil, appears at the terminal of the receiver coil.

$$
\begin{aligned}
v_r(t) &= \frac{d}{dt}\Phi \\
&= M\frac{di_1(t)}{dt} \\
&= M\omega I_1 \cos(\omega t)
\end{aligned}
\tag{1}
$$

In electromagnetic induction systems, the receiver terminal voltage $v_r(t)$ becomes so large that enough power transmission is possible if some load resistance is connected to the terminal to consume properly the transmitted power. This is mainly because the mutual inductance M of coils is so large by the proximity of them and by the iron core that prevents magnetic flux leakage as mentioned above.

Whereas, assuming that the coil pair is set apart, and the magnetic flux generated by the transmitter coil diffuses around it without any spacial constraint, only a fragment of the magnetic flux interlinks the receiver coil and their mutual inductance M becomes small. In this case, only a limited voltage $v_r(t)$ appears at the receiver terminal and insufficient power transmission can be obtained even with the proper load resistance at the terminal.

The magnetic resonant system compensates this problem of small M of the detached coil pair with their magnetic resonance as described in the next subsection.

2.2 Magnetic resonant system

In case that the coil pair is set apart in the electromagnetic induction system shown in Fig. 1, M becomes small and Eq.(1) indicates that the receiver terminal voltage $v_r(t)$ decreases accordingly. However, Eq.(1) also indicates that the decrease of M can be compensated with high frequency ω or large current amplitude I_1 of the transmitter coil.

Commonly, the power supply frequency ω is specified in electromagnetic induction systems used as voltage converters. Whereas, any system that is designed only to transmit power between coils may be driven at an arbitrary high frequency. However, alternating magnetic field at very high frequency possibly generates electromagnetic radiation that dissipates power around the system. This transmission power loss becomes noticeable especially when the distance of coils approaches the wavelength of the electromagnetic radiation.

The magnetic resonant systems utilize the other factor in Eq.(1) that compensates small M between distant coils: transmitter coil current amplitude I_1. This principle is explained by a simple model of a magnetic resonant system shown in Fig. 2 where inductances of coils are set to the same value L for the simplicity and their mutual inductance M is assumed to be small. In this model of the magnetic resonant system, a parallel capacitance C is connected to each coil of the system and therefore a pair of LC-loops are configured that makes the magnetic resonance possible between the loops with huge coil currents[1]. These huge coil currents induced in the magnetic resonance between the LC-loops compensates small M of distant coils providing adequate receiver terminal voltage $v_r(t)$.

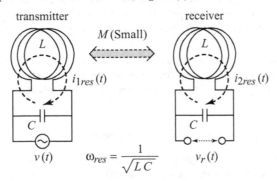

Fig. 2. Magnetic resonance without power transmisson.

The resonance between these LC-loops occurs as follows. First, at some frequency ω, the current $i_1(t)$ of the transmitter coil generates alternating magnetic flux and a segment of this flux interlinks the receiver coil. Second, to decrease this interlinking flux alternation, inverted current $i_2(t)$ begins to flow in the receiver coil and a segment of the inverted flux interlinks the transmitter coil. This negative return of a flux segment decreases the flux alternation inside the transmitter coil mitigating its inductive reactance. Third, because of the decreased

---

[1] In many of the magnetic resonant systems, such as the experimental system reported in (Kurs et al., 2007), each C is not visible as a condenser but exists as a stray capacitance within the coils.

reactance of the transmitter coil that is driven by the stable voltage source $v(t)$, its coil current $i_1(t)$ increases generating more the magnetic flux and the process returns to the first stage [2]. These stages repeats until the system arrives at some stationary state that depends on the loop impedances of both LC-loops. However, when ω approaches the resonance frequency ω_{res} that equals $1/\sqrt{LC}$, because the loop impedance of each LC-loop becomes zero at this critical point, its loop current grows to huge one $i_{1res}(t)$ at the transmitter, or $i_{2res}(t)$ at the receiver as shown in Fig. 2.

This critical incident, where currents of LC-loop pair diverges synchronously at the resonance frequency ω_{res}, is called *magnetic resonance* because their mutual interference is mediated by alternating magnetic field. When this magnetic resonance occurs in the system shown in Fig. 2, the huge coil currents $i_{1res}(t)$ and $i_{2res}(t)$ generates noticeable terminal voltage $v_r(t)$ that is formulated by:

$$v_r(t) = M \frac{d}{dt} i_{1res}(t) + L \frac{d}{dt} i_{2res}(t),\tag{2}$$

or

$$v_r(t) = \frac{1}{C} \int i_{2res}(t)\, dt.\tag{3}$$

Moreover, if proper load resistance is connected to the terminal, enough power becomes available that the transmitter supplies in spite of long distance between the coils.

As described above, a magnetic resonant system utilizes magnetic resonance between LC-loops of transmitter and receiver. The role of the magnetic resonance is to utilize huge coil currents that makes the power transmission possible between the loops even though they are located apart. According to Eq.(1), this principle of magnetic resonant system can be explained that the huge coil current $i_1(t)$ compensates the small M to generate a sufficient amplitude of $v_r(t)$.

In the next section, the performance of power transmission of a magnetic resonant system is investigated with a load resistance at the receiver. Other characteristics are also examined including actual resonance frequency ω_{res} and amplitudes of resonance currents $i_{1res}(t)$ and $i_{2res}(t)$.

3. Power transmission by magnetic resonant system

In magnetic resonant systems, voltage source supplies electric power to the LC-loop at the transmitter and load resistance at the receiver extracts this power from the LC-loop of its side. Between the LC-loops, alternating magnetic field carries the power at the resonance frequency where both coil currents diverge synchronously fastening their magnetic coupling.

In this section, power transmission characteristics of a magnetic resonant system with a load resistance is investigated. As the result, examples are shown of actual resonance frequency, resonance currents of coils, and transmitted power in relation with system parameters.

---

[2] This intuitive description of the system behavior focusing on the inductive reactances of LC-loops is valid when ω exceeds the resonance point of the loops. Because the capacitive reactances become dominant in the loops when ω falls behind the resonance point, $i_1(t)$ and $i_2(t)$ become inphase in this range of ω and the description must be modified accordingly.

3.1 System analysis

Figure 3 shows a magnetic resonant system adopted for the investigation. This system is equivalent to that shown before in Fig. 2 except for the load resistance R of the receiver and notations of currents $i_1(t) \sim i_4(t)$ each for the indicated line in the figure. Magnetic coupling between coils is indicated by mutual inductance M or coupling factor k. k equals M/L and therefore never exceeds unity. The power supplied by the voltage source $v(t)$ (input power) and that consumed by the load resistance R (output power) are denoted by P_{in} and P_{out}, respectively.

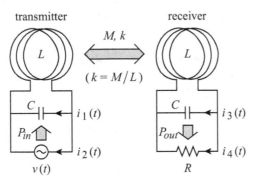

Fig. 3. Power transmission by magnetic resonant system.

Assuming that the system is at some stationary state, the voltage source $v(t)$ is expressed by complex exponential as $V \exp(j\omega t)$ and $i_1(t) \sim i_4(t)$ are expressed as $I_1 \exp(j\omega t) \sim I_4 \exp(j\omega t)$, respectively. Voltage amplitude V is set to be a real number. Whereas, the amplitudes of currents $I_1 \sim I_4$ may be complex numbers indicating the phase shift of $i_1(t) \sim i_4(t)$ from $v(t)$, respectively.

Among these amplitudes and parameters of the system shown in Fig. 3, simultaneous equations:

$$\begin{cases} V = j\omega \left\{ L(I_1 + I_2) + M(I_3 + I_4) \right\} \\ V = -I_1 / (j\omega C) \\ 0 = R\,I_4 + j\omega \left\{ L(I_3 + I_4) + M(I_1 + I_2) \right\} \\ R\,I_4 = I_3 / (j\omega C) \end{cases} \tag{4}$$

are established. Here, the angular frequency ω is assumed to be low enough in relation to the coil distance. Therefore, in the first and third lines of the equations, the interaction between coils by M occurs with no phase difference[3].

The input power P_{in} and output power P_{out} are derived as:

$$P_{in} = \frac{1}{2} V \Re(I_2), \quad P_{out} = \frac{1}{2} R\, |I_4|^2. \tag{5}$$

($\Re(x)$ means the real part of x.) These values must be equal to each other according to energy conservation law. This is because power dissipation caused by electromagnetic radiation or

---

[3] For example, in the case of $\omega = 2\pi \cdot 1[\text{MHz}]$ corresponding to the electromagnetic wave length of 300[m], coil distance of 3[m] or less may satisfy this assumption.

other power consumers such as inner resistances of coils are not included in Fig. 3 nor in Eq.(4). This is confirmed by calculating Eq.(5) with I_2 and I_4 derived from Eq.(4). As the result, the equivalent transmission power P is obtained as[4],

$$P = P_{in} = P_{out} = \frac{1}{2} \cdot \frac{V^2 k^2 R}{L^2(1-k^2)^2 \omega^2 + \{1 - CL(1-k^2)\omega^2\}^2 R^2}. \tag{6}$$

This expression of power P is valid on the system shown in Fig 3 that is driven at any frequency ω. However, the system becomes magnetic resonant system only at the resonance frequency ω_{res}. In the next subsection, this frequency ω_{res} is examined.

3.2 Resonance frequency

The resonance frequency ω_{res} is derived here assuming that the transmission power P becomes maximum at this critical frequency. According to this assumption, ω_{res} and the maximum power P_{res} is determined as[5]:

$$\omega_{res} = \sqrt{\frac{2CR^2 - L(1-k^2)}{2L(1-k^2)C^2 R^2}}. \tag{7}$$

$$P_{res} = \frac{2V\, 2k^2 C^2 R^3}{L(1-k^2)\,(4CR^2 - L(1-k^2))} \tag{8}$$

Besides these expressions, that are rather complicated and cannot be recognized intuitively, approximated ones $\tilde{\omega}_{res}$ and $\tilde{P}_{res}$ are available. First, $\tilde{\omega}_{res}$ is:

$$\omega_{res} \simeq \tilde{\omega}_{res} = \frac{1}{\sqrt{LC(1-k^2)}}. \tag{9}$$

This approximation is easily recognized by the equation:

$$\omega_{res} = \tilde{\omega}_{res} \cdot \sqrt{1 - \frac{\epsilon}{2}}, \quad \epsilon = \frac{1}{RC\tilde{\omega}_{res}}. \tag{10}$$

As explained later, ϵ becomes very small provided the system is designed to perform as a magnetic resonant system. Therefore, in most cases, $\tilde{\omega}_{res}$ in Eq.(9) approximates well the resonance frequency ω_{res}.

The approximated form $\tilde{\omega}_{res}$ indicates that the resonance frequency of the system shown in Fig. 3 shifts from the inherent point $1/\sqrt{LC}$ of a LC-loop depending on the coupling factor k of the coil pair. k falls to 0 when the coils are located far away and approaches 1 when they become close to each other. Therefore, the resonance frequency $\tilde{\omega}_{res}$ increases from the inherent point when the distance of LC-loops are set properly so as to they can perform power transmission.

The variable ϵ in Eq.(10) indicates the degree of how well the RC parallel impedance can be approximated to only resistance R at the frequency $\tilde{\omega}_{res}$. This RC parallel impedance is connected to the terminal of receiver coil shown in Fig. 3. The variable ϵ equals the ratio of the absolute value of impedance $1/(j\omega C)$ of the capacitance C to the resistance R.

---

[4] The solutions $I_1 \sim I_4$ of Eq.(4) are written in Appendix A.1.
[5] The derivation process of Eq.(7) and (8) is described in Appendix A.2.

A small ϵ means that the capacitive impedance dominates in the RC parallel impedance and therefore the receiver forms almost pure LC-loop[6]. Because this LC-loop must work as a resonator in magnetic resonant systems, R must be set properly to make ϵ small. Whereas, in case that the resistance R is set small in relation to the capacitive impedance $1/(j\omega C)$, the circuit almost becomes resistance R. In this case, the LC-loop of the receiver scarcely work as a resonator to receive power from the transmitter. Furthermore, when R falls behind a threshold, ω_{res} in Eq.(10) becomes imaginary quantity. In such cases, the magnetic resonance cannot occur. This threshold R_{th} is derived by $\epsilon = 2$ as,

$$R_{th} = \frac{1}{2C\widetilde{\omega}_{res}} = \frac{1}{2}\sqrt{\frac{L(1-k^2)}{C}}. \tag{11}$$

This means that if receiver load resistance R adequately exceeds this value R_{th}, the system shown in Fig. 3 performs well as a magnetic resonant system and the approximation $\widetilde{\omega}_{res}$ of the resonance frequency ω_{res} is valid. Whereas, if R is set equal to or less than R_{th}, the system cannot be a magnetic resonant system at any frequency ω.

Finally, the approximated value of the maximum power transmission $\widetilde{P}_{res}$ is derived by substitution of $\widetilde{\omega}_{res}$ (Eq.(9)) into P (Eq.(6)) as:

$$\widetilde{P}_{res} = \frac{V^2k^2CR}{2L(1-k^2)}. \tag{12}$$

3.3 An example of power transmission

Based on the equations derived in the previous subsections, an example of power transmission characteristics of a magnetic resonant system is shown with actual parameters. Besides the coil inductance L, other parameters of coil pair are assumed to be the same. The parameters are set as follows:

Coil radius a, its wire turns N, and its thickness t are set to 5[cm], 20[turns], and 5[mm], respectively. From these values, its self inductance L is calculated to be 97.9[μH][7].

The inherent resonance frequency of the LC-loop is set to 10[KHz]. This value is not the angular frequency but the actual frequency and equals $1/(2\pi\sqrt{LC})$. Because L is already fixed, the capacitance C is derived from this frequency as 2.59[μF].

Eq.(11) indicates that the threshold R_{th} of the load resistance R becomes maximum when k equals 0. This maximum R_{th} is derived from L and C as 3.07[Ω]. As an adequately exceeding value over this R_{th}, the load resistance R of the receiver is set to 500[Ω]. These parameters are listed in Table 1.

With these parameters adopted, the resonance frequency f_{res} of the magnetic resonant system and coupling factor k of its coil pair are calculated and shown in Fig. 4. Horizontal axis represents the distance d between the coils. f_{res} is calculated by Eq.(9). k is calculated from coil parameters and the distance d assuming that coils are confronting each other with the same axis[7]. As this graph shows, both values increase as the distance d of coils decreases. However, the increase of the resonance frequency f_{res} from the inherent value $1/(2\pi\sqrt{LC})$ is not noticeable until d falls behind the diameter 10[cm] of coils.

---

[6] Notice that a small impedance dominates in parallel circuit because electric current tends to flow the wire with smaller impedance than others.

[7] The calculations of L and k are described in Appendix A.3.

coil parameters	notation	value	unit
inductance	L	97.9	μH
radius	a	5	cm
wire turn	N	20	turn
thickness	t	5	mm
other parameters	notation	value	unit
capacitance	C	2.59	μF
load resistance	R	500	Ω
voltage amplitude	V	100	Volt

Table 1. System parameters.

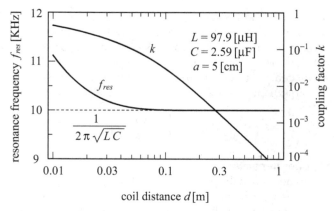

Fig. 4. Resonance frequency f_{res} of the system and coupling factor k of the coil pair.

When the voltage source $v(t)$ in Fig. 3 drives the system at the resonance frequency ω_{res} (or f_{res} shown in Fig. 4), magnetic resonance occurs between its LC-loops with huge loop currents and with power transmission from the transmitter to the receiver. This transmission power P_{res} and loop currents of the magnetic resonant system are calculated and shown in Fig. 5. Horizontal axis represents coil distance d as same as Fig. 4. Vertical axis represents P_{res} and amplitudes $|I_1| \sim |I_4|$ of the electric currents $i_1(t) \sim i_4(t)$, respectively. In addition to the parameters specified above, the amplitude V of the voltage source $v(t)$ is set to 100[V]. P_{res} is calculated by Eq.(12) and amplitudes of the electric currents are calculated by Eq.(A1)~(A4) in Appendix.

This graph shows several features on this magnetic resonant system as follows:

- The transmitted power P_{res} varies noticeably from about 650[W] to 0.01[W] as the distance d of coils increases from their radius 5[cm] to its ten times long 50[cm].

- Whereas, the amplitudes of resonance loop currents $|I_1|$ and $|I_3|$ do not decrease similarly. Especially, $|I_1|$ keeps almost constant value 16[A] throughout the indicated range of d.

- In comparison with these huge resonance currents, $i_2(t)$ and $i_4(t)$ perform actual power transmission (see Eq.(5)). Concerning the practical distance d of coils over their diameter 10[cm], the amplitudes of these currents $|I_2|$ and $|I_4|$ do not exceed moderate value.

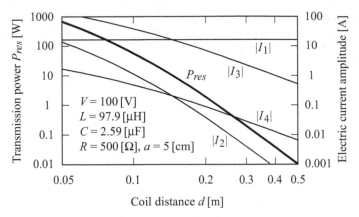

Fig. 5. Transmission power P_{res} and current amplitudes.

The noticeable decrease of transmission power P_{res} along with the increase of coil distance d seems not to be a serious problem according to Eq.(12). This equation indicates that the decrease of P_{res} can be compensated with the square of voltage amplitude V.

On the other hand, huge amplitudes of loop currents $|I_1|$ and $|I_3|$ possibly affect seriously the power transmission efficiency of the system. This is because the inner resistances of coils consume a part of transmission power in proportional to the square of current amplitude $|I_1|$ or $|I_3|$.

In the next section, these power losses caused by inner resistances of coils are taken into account on the analysis of magnetic resonant system performance. As the results, the designs are obtained for the power transmission efficiency maximum or for the effective power supply maximum. In addition, it is found that any combination of the efficiency and power supply is possible provided the system is driven over a critical frequency.

4. Optimal designs of magnetic resonant system with internal resistances of coils

As shown in the previous section, example values of coil currents ($|I_1|$ and $|I_3|$ in Fig. 5) become huge at the resonance point of the system in comparison with other currents ($|I_2|$ and $|I_4|$ in the same figure) that concern actual power transmission.

These huge coil currents may bring noticeable power dissipation caused by internal resistances of coils even though their amounts are insignificant. This is because the Joule loss within a wire of resistance R is proportional to the square of its current flow $i(t)$ as $Ri(t)^2$ and therefore is sensitive to the current amplitude especially when the amplitude is very large.

Therefore, in actual designs of magnetic resonant systems, internal resistances of coils must be taken into account and appropriate system parameters must be determined from the viewpoint of the power transmission performances that may degrade caused by the internal resistances of coils.

In this section, power transmission performances of magnetic resonant systems are investigated taking into account the internal resistances of coils. As the results, appropriate system designs are obtained for the optimal system performances.

4.1 System analysis

Figure 6 shows the magnetic resonant system with the internal resistances of coils adopted for the investigation. The basics of the system is equivalent to that shown in Fig. 3. The differences from this previous system are as follows:

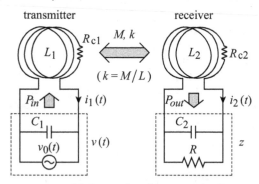

Fig. 6. Magnetic resonant system with internal resistances of coils.

1. For the generality of investigations and system designs, constraint of symmetry between LC-loops of transmitter and receiver is eliminated from their parameters.

2. According to this scheme of generality, coil inductances of transmitter and receiver are set to L_1 and L_2, respectively. (From now on, parameters with suffix '1' means that of transmitter and suffix '2' means that of receiver.) Similarly, capacitances are set to C_1 and C_2.

3. For the evaluation of Joule losses within the coils, the internal resistances R_{c1} and R_{c2} are additionally adopted as shown in Fig. 6.

4. $i_1(t)$ and $i_2(t)$ in Fig. 3 are integrated into unique coil current $i_1(t)$ of the transmitter. Similarly, coil currents of receiver are integrated into $i_2(t)$.

In addition, some elements are substituted by integrated ones as follows for the convenience of the analysis.

1. The part of the transmitter consists of C_1 and the voltage source (denoted by $v_0(t)$) is substituted by a single voltage source $v(t)$ as shown in the left part of Fig. 6. This substitution is possible because the voltage and the frequency with that the transmitter coil is driven is determined by $v_0(t)$ itself. Whereas, C_1 only let the resonance current $i_1(t)$ bypass this voltage source.

2. The part of the receiver consists of C_2 and R is substituted by a complex impedance z as shown in the right part of Fig. 6. This substitution embeds C_2 in z as a component of capacitive reactance.

According to the same assumption as put in the previous section, that the system is at some stationary state, the voltage source $v(t)$ is expressed by $V \exp(j\omega t)$. Similarly, $i_1(t)$ and $i_2(t)$ are expressed by $I_1 \exp(j\omega t)$ and $I_2 \exp(j\omega t)$, respectively. The amplitudes of currents I_1 and I_2 may be complex numbers indicating the phase shift of $i_1(t)$ and $i_2(t)$ from $v(t)$, respectively. Among these amplitudes and parameters of the system shown in Fig. 6, simultaneous equations:

$$\begin{cases} V = j\omega L_1 I_1 + j\omega M I_2 + R_{c1} I_1 \\ z I_2 + j\omega L_2 I_2 + j\omega M I_1 + R_{c2} I_2 = 0 \end{cases} \tag{13}$$

are established[8]. The power supply P_{in} from $v(t)$, and the power consumption P_{out} of z are given by,

$$P_{in} = \frac{1}{2} V \Re(I_1), \quad P_{out} = \frac{1}{2} \Re(z) \, | \, I_2 \, |^2. \tag{14}$$

In the previous section, it was confirmed by Eq.(6) that P_{in} equals P_{out}. However, because of internal resistances of coils R_{c1} and R_{c2} that consume a part of P_{in} at the transmitter and that of P_{out} at the receiver, P_{in} and P_{out} become different to each other in the system shown in Fig. 6. Taking into account these power dissipations of Joule losses by R_{c1} and R_{c2}, the relational expression:

$$P_{out} = P_{in} - \frac{1}{2} R_{c1} \, |I_1|^2 - \frac{1}{2} R_{c2} \, |I_2|^2 \tag{15}$$

is established[9] (from now on, P_{out} is called as *effective power supply*). Eq.(15) indicates that the power transmission efficiency:

$$\xi = \frac{P_{out}}{P_{in}} \tag{16}$$

exists somewhere below one depending on the system parameters and the resonance frequency that the system is driven at. (from now on, this efficiency ξ is simply called as *transmission efficiency*.)

In the following subsections, first, the internal resistance of a coil is estimated in relation with the coil configuration. Based on this estimation, second, the optimal receiver impedance z is investigated that maximizes the transmission efficiency ξ or effective power supply P_{out} itself. Finally, examples of these optimal designs are shown with actual parameters of magnetic resonant systems.

4.2 Internal resistance of coils

According to Eq.(15), internal resistances R_{c1} and R_{c2} must be reduced enough to obtain admissible transmission efficiency ξ of the magnetic resonant system. An elementary way to reduce the internal resistances is to extend the cross section of wire of the coils. This is because the internal resistance of a wire is inversely proportional to its cross section.

However, unrestrained increase of wire radius causes enlargement of the coil configuration. In many designs of magnetic resonant systems, the dimensional specifications exist that restrict the coil configuration. Because of this reason, in the estimation of internal resistance of a coil, the coil configuration must be specified first, and then the internal resistance of the coil is derived in relation with its wire turns, coil inductance, and maximum wire radius under the restriction of the coil configuration.

---

[8] As mentioned in the previous section, the angular frequency ω is assumed to be low enough in relation to the coil distance.

[9] When a current $i(t)$ is expressed by $I \exp(j\omega t)$, Joule loss $R\,i(t)^2$ is estimated as $(1/2)R|I|^2$ in average.

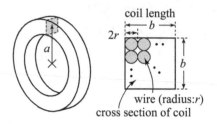

Fig. 7. Coil configuration.

Figure 7 shows the coil configuration adopted to the estimation of the internal resistance. It is assumed that the cross section of coil is square and the wire is wound densely within the cross section. Parameters a, b, and r represent coil radius, coil length, and wire radius, respectively. According to these assumptions and notations, equations:

$$\begin{cases} L = 2 \cdot 10^{-7} \pi^2 K(\eta) a N^2 / \eta \\ R_c = \rho \cdot 2\pi a N / (\pi r^2) \end{cases} \tag{17}$$

are obtained where L and R_c represent coil inductance and its internal resistance, respectively. K, η, N, and ρ are Nagaoka coefficient, configuration index, wire turns, and the resistivity of wire material, respectively[10]. Configuration index η equals $b/2a$ and wire turns N approximately equals $(d/2r)^2$.

From these equations, a simple relation between the coil inductance L and its internal resistance R_c is derived as:

$$R_c = \sigma L, \ \sigma = \frac{10^7}{\pi^2} \cdot \frac{\rho}{a^2 K(\eta) \eta} . \tag{18}$$

where σ denotes a proportional coefficient between R_c and L and has unit of $(1/\text{second})$[11]. Because the expression of σ does not include wire parameters N and r, this coefficient is determined only by the coil configuration and the resistivity ρ of wire material.

4.3 Optimal performances of magnetic resonant system
For the optimal performances of magnetic resonant system, two types of receiver impedance z are derived.

1. Optimal impedance z^e that maximizes the transmission efficiency ζ provided other parameters of the system are fixed.

2. Optimal impedance z^p that maximizes the effective power supply P_{out} provided other parameters of the system are fixed.

---

[10] The second equation of (17) assumes that R_c inversely proportional to the wire cross section (πr^2). However, this simple assumption will not valid when the coil current alternates at high frequency because of the skin effect of wire(Maqnusson, et al.). Therefore, Eq.(17) must additionally assume that the frequency is low enough or the wire is *Litz wire* that reduces the skin effect(Kazimierczuk, 2009).

[11] Considering that ωL and R_c have the same unit (Ω), σ must have the same unit as ω according to Eq.(18).

The process of derivation of z^e and z^p is as follows. First, I_1 and I_2 are derived from Eq.(13). Second, the effective power supply P_{out} and the transmission efficiency ξ are derived from Eq.(14) and (16), respectively. Third, z^e and the maximum transmission efficiency ξ_m are derived along with z^p and the maximum effective power supply P_m (from now on, simply called as *maximum power*). Finally, according to Eq.(18), equations $R_{c1} = \sigma_1 L_1$ and $R_{c2} = \sigma_2 L_2$ are applied to these results for the elimination of internal parameters[12].
The derived formulas are as follows. First, the maximum transmission efficiency ξ_m, the real part z_r^e and the imaginary part z_i^e of z^e are

$$\xi_m = 1 - 2 \frac{\sqrt{\sigma_1\sigma_2}\sqrt{\sigma_1\sigma_2 + k^2\omega^2} - \sigma_1\sigma_2}{k^2\omega^2} \tag{19}$$

$$z_r^e = L_2 \sqrt{\sigma_2^2 + \frac{k^2\sigma_2\omega^2}{\sigma_1}}, \quad z_i^e = -L_2\omega \tag{20}$$

Obviously, transmission efficiency does not depend on the voltage source amplitude V. This is confirmed by (19). In addition, this formula is independent also of inductances L_1 and L_2. These coil parameters are hidden in σ_1 and σ_2 that specifies the coil configurations.
Second, the maximum power P_m, the real part z_r^p and the imaginary part z_i^p of z^p are

$$P_m = \frac{V^2 k^2 \omega^2}{8L_1 \left(\sigma_1^2\sigma_2 + k^2\sigma_1\omega^2 + \sigma_2\omega^2\right)} \tag{21}$$

$$z_r^p = L_2 \frac{\sigma_1^2\sigma_2 + (\sigma_1 k^2 + \sigma_2)\,\omega^2}{\sigma_1^2 + \omega^2}, \quad z_i^p = -L_2\omega \frac{\sigma_1^2 + (1 - k^2)\,\omega^2}{\sigma_1^2 + \omega^2} \tag{22}$$

In contrast to Eq.(19), Eq.(21) includes V and L_1. However, as the essential value of maximum power, normalized maximum power $\widehat{P_m}$ is derived independently of V and L as:

$$\widehat{P_m} = (L_1/V^2)P_m = \frac{k^2\omega^2}{8\left(\sigma_1^2\sigma_2 + (\sigma_1 k^2 + \sigma_2)\omega^2\right)} \tag{23}$$

This normalized maximum power $\widehat{P_m}$ has unit of (second)[13] and is utilized in the process of optimal system designs for the maximum transmission efficiency described in the next subsection.

4.4 Examples of optimal system design
According to the formulas derived in the previous subsection, the system design procedure for the optimal performance of magnetic resonant systems is shown with actual design examples. Though the previous analyses can be applied to the cases where transmitter and receiver coils are not symmetrical, they are assumed to be symmetrical in the followings. The symmetrical coil pair is essential in a power supply link between mobile robots that configure power supply network of multi-robot systems(Sugiyama, 2009; 2010; 2011). Because of this reason, coil inductances and their internal resistances are substituted by the same variable L and R_c, respectively.

---

[12] The details of this process are described in the Appendix A.4~A.7.
[13] According to Eq.(23), and considering that k has no unit, the unit of $\widehat{P_m}$ is the same as (ω^2/σ^3) and is equal to $(1/\sigma)$.

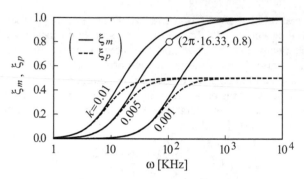

Fig. 8. Maximum transmission efficiency ξ_m when $z = z^e$ and transmission efficiency ξ_p when $z = z^p$.

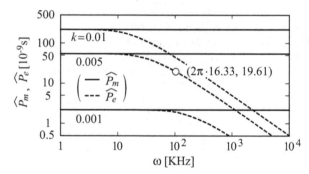

Fig. 9. Normalized maximum power $\widehat{P_m}$ when $z = z^p$ and normalized power $\widehat{P_e}$ when $z = z^e$.

Figure 8 indicates the maximum transmission efficiency ξ_m versus frequency ω by solid lines calculated by Eq.(19). Dashed lines indicate the transmission efficiency ξ_p when the impedance is set as z^p that maximizes effective power supply.

On the other hand, Fig. 9 indicates the normalized maximum power $\widehat{P_m}$ versus frequency ω by solid lines calculated by Eq.(23). Dashed lines indicate the normalized power $\widehat{P_e}$ that is derived by multiplying P_e by (L/V^2) similarly to $\widehat{P_m}$. Where, P_e equals effective power supply P_{out} when the receiver impedance is set as z^e that maximizes transmission efficiency. ξ_p and $\widehat{P_e}$ are expressed as follows:

$$\xi_p = \frac{k^2\omega^2}{2\left(2\sigma^2 + k^2\omega^2\right)} \tag{24}$$

$$\widehat{P_e} = (L_1/V^2)\, P_e = \frac{k^2\omega^2\sqrt{\sigma^2 + k^2\omega^2}}{2\left(\sigma^2 + \omega^2 + k^2\omega^2\right)\left(2\sigma^2 + k^2\omega^2 + 2\sigma\sqrt{\sigma^2 + k^2\omega^2}\right)} \tag{25}$$

In both figures, three values of coupling factor k between coils are applied as indicated. Other parameters are as follows:

The wire material of coils is assumed to be copper. Therefore, its resistivity ρ is set to $1.72 \cdot 10^{-8}[\Omega\,\text{m}]$. The coil radius a and its configuration index η are set to 5[cm] and 0.3, respectively.

Applying these values to Eq.(18), the proportional coefficient σ between R_c and L is calculated to be 57.37[1/s]. These parameters are listed in Table 2 [14].

coil parameters	notation	value	unit
wire resistivity	ρ	$1.72 \cdot 10^{-8}$	Ωm
radius	a	5	cm
configuration index	η	0.3	-
proportional coefficient	σ	57.37	1/s

Table 2. System parameters.

Figure 8 shows two points. First, when the coupling factor k is fixed according to the relative position of the coils, the minimum frequency ω_{min} exists, which satisfies the target value ζ_{target} of the transmission efficiency. Second, when the impedance z^p is adopted for maximum power supply, ζ_p cannot exceed 50% even if the frequency ω increases enough. Therefore, in the practical designs where ζ_{target} exceeds 50%, z^p cannot be applied as the receiver impedance. On the other hand, Fig. 9 shows that when $z = z^p$, the normalized maximum power $\widehat{P_m}$ stays nearly constant, whereas when $z = z^e$, the normalized power $\widehat{P_e}$ decreases as ω increases. However, because the normalized value $\widehat{P_e}$ is derived from P_e multiplied by (L/V^2) as shown by Eq.(25), any amount of power supply is possible by adjusting this coefficient.

From these results, a design procedure of a magnetic resonant system for the maximum transmission efficiency is derived as follows.

1. Specify these items:
   - System parameters except for the receiver impedance z.
   - Target transmission efficiency ζ_{target} that exceeds 50%.
   - Target effective power supply P_{target}.

2. According to system parameters except for z, and to ζ_{target}, the minimum frequency ω_{min} is determined.

3. To make minimum the power dissipation by electromagnetic radiation, the frequency ω of the voltage source $v_0(t)$ must be as low as possible. Because of this reason, ω is fixed to the minimum frequency ω_{min}.

4. At this frequency ω_{min}, the value of normalized power $\widehat{P_e}$ is determined. Then, (V^2/L) is derived from $P_{target}/\widehat{P_e}$.

5. Under the constraint that (V^2/L) is fixed, voltage source amplitude V and coil inductance L can be set arbitrary. However, in many cases L may be set first because it affects receiver impedance z^e.

6. The optimal receiver impedance z^e is determined by Eq.(20) that makes the magnetic resonant system satisfy the target efficiency ζ_{target} and target effective power supply P_{target} at the minimum frequency ω_{min}.

According to this design procedure, a design example is described as follows:

---

[14] At these parameters, the minimum value 0.001 of k corresponds to the coil distance of 44.7[cm] (the relationship between $k(= M/L)$ and the coil distance d is described in Appendix A.3). On the other hand, the maximum value 10^7 [Hz] of the indicated range of ω corresponds to the wavelength of 190[m]. Therefore, the assumption put on Eq.(13) may be valid over these figures.

1. System parameters are specified by Table 2. In addition, coupling factor k of coils is set to 0.005 (the distance d between them is 25.5[cm][15]). ζ_{target} is set to 80% and P_{target} is set to 10[W].

2. ω_{min} is found to be $2\pi \cdot 16.33$[KHz] from Fig. 8 (a white dot indicates this point).

3. $\widehat{P_e}$ is found to be 19.61×10^{-9}[s] from Fig. 9 (a white dot indicates this point).

4. P_{target} (10[W]) is obtained when $(V^2/L) = 5.100 \times 10^8$. Under this constraint, L and V are fixed to 0.3331[mH] and 412.2[V], respectively[16].

5. According to Eq.(20), z^e is obtained as $0.1720 - j34.18$. This impedance corresponds to $R = 6.795$[KΩ] and $C = 0.2850[\mu F]$ for RC-parallel impedance as shown in Fig. 6.

Besides this design example of magnetic resonant system, more general design specifications are indicated in Fig. 10 and 11 with extended range of system parameters.

Figure 10 indicates voltage source amplitude V by solid lines and its minimum frequency f_{min} by dashed lines. Whereas Fig. 11 indicates receiver load resistance R by solid lines and LC-loop capacitance C by dashed lines.

In both figures, system parameters and target power supply P_{target} are the same as the previous system design. Coupling factor k of coils is set to three different values: 0.001, 0.005, 0.01 as shown in the figure. These values correspond to 44.7, 25.5, and 19.8[cm] of coil distance d, respectively. Horizontal axis represents ζ_{target} that exceeds 0.5.

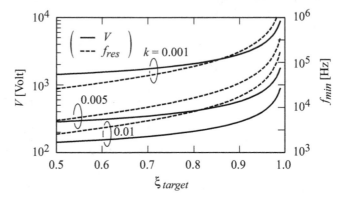

Fig. 10. Voltage source amplitude V and minimum frequency f_{min}.

Fig. 10 shows the difficulty of system design when coils are set apart. For example, when k equals 0.001, this corresponds to 44.7[cm] of coil distance d as mentioned above, voltage source amplitude V exceeds 1000[V] at 0.7 of ζ_{target}. This impractical voltage amplitude may be reduced by adjusting L under the constraint of (V^2/L) is fixed. However, because L affects the optimal receiver impedance z^e concerning actual configuration of RC parallel impedance, reasonable value of L must be investigated further in relation with the whole system design. The influence of L to the RC parallel impedance can be recognized in Fig. 11. When ζ_{target} increases in Fig. 10, the frequency f_{min} increases especially at high ζ_{target}. This means that the inherent resonance frequency $1/\sqrt{LC}$ increases accordingly at high ζ_{target}. However,

---

[15] d can be inversely derived from k by Eq.(A9) and $k = L/M$.
[16] This value of L is derived from Eq.(17) substituting 50 for the wire turn N.

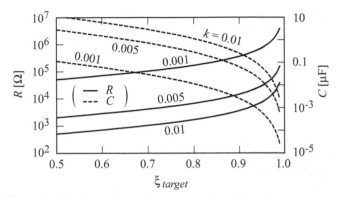

Fig. 11. Receiver load resistance R and LC-loop capacitance C.

because L is fixed to the constant value in this case, C must decrease excessively as shown in Fig. 11. This small value of C constrains R to increase enough to keep its domination in RC parallel impedance and to keep the receiver approximately pure LC-loop as discussed before in subsection 3.2. Because of this reason, the values of R and C alternate excessively when ζ_{target} is high. Adjustment of L to some reasonable value will alleviate this excessive alternation of the system parameters.

5. Conclusion

Performances of wireless resonant energy links based on nonradiative magnetic field generated by a pair of coils (magnetic resonant system) are analyzed using electrical circuit theory.

First, the difference of principle of magnetic resonant systems from that of conventional electromagnetic induction systems is explained. As the result, it became clear that the role of magnetic resonance is to keep the magnetic coupling between coils that are located apart by compensating their small mutual inductance with huge coil currents.

Second, based on the difference from the conventional systems, the inherent characteristics of power transmission of magnetic resonant systems are explained including the expressions of resonance frequency and power transmission at the resonance frequency. In addition, examples of huge resonance currents are indicated that may degrade power transmission efficiency with internal resistances of coils.

Finally, taking into account the internal resistances of coils, optimal system designs are investigated that maximize the power transmission efficiency or effective power supply under constraints of coil sizes and voltages of power sources. As the result, the minimum resonance frequency is determined where the magnetic resonant system satisfies both of target transmission efficiency and target power supply provided the receiver impedance is set properly.

In further studies, these results must be confirmed by comparisons with ones derived from experimental systems. The comparison with experimental results by MIT(Kurs et al., 2007) is already reported(Sugiyama, 2009) where the possibility is shown that the transmission efficiency of the experimental system can be improved with the optimal designs derived in Section 4. In addition, characteristics of an experimental system with symmetrical coil pair is reported(Sugiyama, 2011) that confirms the theoretical results. However, characteristics of actual magnetic resonant systems with more general designs such as asymmetrical coil pair must be examined to confirm the theoretical results.

6. Appendix[17]

A.1 Expressions of current amplitudes in Fig. 3

$$I_1 = -jVC\omega \tag{A1}$$

$$
I_2 = \frac{k^2 RV}{(1-k^2)^2 L^2 \omega^2 + (R - C(1-k^2)LR\omega^2)^2}
$$

$$
+ j\, \frac{V\left(1 - C(1-k^2)L\omega^2\right)\left(R^2\left(CL\omega^2\left(2 - C(1-k^2)L\omega^2\right) - 1\right) - (1-k^2)L^2\omega^2\right)}{L\omega\left((1-k^2)^2 L^2 \omega^2 + (R - C(1-k^2)LR\omega^2)^2\right)} \tag{A2}
$$

$$
I_3 = \frac{-Ck(1-k^2)LRV\omega^2}{(1-k^2)^2 L^2 \omega^2 + (R - C(1-k^2)LR\omega^2)^2}
$$

$$
- j\, \frac{CkR^2 V\omega\left(1 - C(1-k^2)L\omega^2\right)}{(1-k^2)^2 L^2 \omega^2 + (R - C(1-k^2)LR\omega^2)^2} \tag{A3}
$$

$$
I_4 = \frac{kRV\left(C(1-k^2)L\omega^2 - 1\right)}{(1-k^2)^2 L^2 \omega^2 + (R - C(1-k^2)LR\omega^2)^2}
$$

$$
+ j\, \frac{k(1-k^2)LV\omega}{(1-k^2)^2 L^2\,\omega^2 + (R - C(1-k^2)LR\omega^2)^2} \tag{A4}
$$

A.2 Derivation of Eq.(7) and (8)
First, transform the Eq.(6) into:

$$
P = \frac{1}{\alpha + \beta\chi + \gamma\chi^2}. \tag{A5}
$$

$$
\begin{pmatrix}
\chi = \omega^2 \\[4pt]
\alpha = \dfrac{2R}{V^2 k^2} \\[8pt]
\beta = \dfrac{2L(1-k^2)\left\{L(1-k^2) - 2CR^2\right\}}{V^2 k^2 R^2} \\[8pt]
\gamma = \dfrac{2L^2 C^2 R^2 (1-k^2)^2 (1+k^2)^2}{V^2 k^2}
\end{pmatrix}
$$

At the maximum point of P, the differential of Eq.(A5) by χ equals zero. Therefore,

---

[17] Many of the expressions below are derived in assistance with Mathematica 6.0.

$$\frac{d}{d\chi}P = -\frac{\beta + 2\gamma\chi}{(\alpha + \beta\chi + \gamma\chi^2)^2} = 0$$

$$\chi = \omega^2 = \frac{-\beta}{2\gamma} = \frac{2CR^2 - L(1 - k^2)}{2L(1 - k^2)C^2R^2} \tag{A6}$$

Eq.(7) equals the square root of Eq.(A6). Eq.(8) is derived by substitution of (7) into (6).

A.3 Derivation of self inductance L and coupling factor k

The self inductance L of a coil is calculated by the equation:

$$L = 2\pi \cdot 10^{-7} \cdot K(rN^2/\eta), \tag{A7}$$

where, η equals (t/a) and K means Nagaoka coefficient(Hayt, 1989).
The coupling factor k equals M/L. Generally, the mutual inductance M of coils C1 and C2 is calculated by:

$$M = 10^{-7} \oint_{c1} \oint_{c2} \frac{d\mathbf{s}_1 \cdot d\mathbf{s}_2}{|\mathbf{r}_1 - \mathbf{r}_2|} \tag{A8}$$

where, $c1$, $\mathbf{r}_1$, and $\mathbf{s}_1$ means the contour integral of coil C1, a vector from the origin of coordinates to a point on C1, and a tangent vector at the point, respectively. Variables with suffix 2 means that of coil C2(Hayt, 1989).
Especially, when the coils have the same radius a and wire turns N, and when they confronting each other with the same axis, a reduced form is available as:

$$M = 4\pi 10^{-7} a N^2 \int_0^\pi \frac{\cos\theta}{\sqrt{2(1 - \cos\theta) + (d/a)^2}} d\theta. \tag{A9}$$

A.4 Expressions of current amplitudes in Fig. 6

$$I_{1r} = \left[V\left(L_1 L_2 k^2 (r_2 + z_r)\omega^2 + r_1\left((r_2 + z_r)^2 + (z_i + L_2\omega)^2 \right) \right) \right] \Big/ D \tag{A10}$$

$$I_{1i} = \left[-V L_1 \omega \left((r_2 + z_r)^2 + (z_i + L_2\omega)\left(z_i + (1 - k^2)L_2\omega \right) \right) \right] \Big/ D \tag{A11}$$

$$I_{2r} = \left[-V\sqrt{L_1 L_2} k\omega (L_1(r_2 + z_r)\omega + r_1(z_i + L_2\omega)) \right] \Big/ D \tag{A12}$$

$$I_{2i} = \left[-V\sqrt{L_1 L_2} k\omega \left(r_1(r_2 + z_r) + L_1\omega \left(-z_i - (1 - k^2)L_2\omega \right) \right) \right] \Big/ D \tag{A13}$$

$$D = 2L_1 L_2 k^2 r_1 (r_2 + z_r)\omega^2 + r_1^2 \left((r_2 + z_r)^2 + (z_i + L_2\omega)^2 \right)$$
$$+ L_1^2\omega^2 \left((r_2 + z_r)^2 + \left(z_i + (1 - k^2)L_2\omega \right)^2 \right) \tag{A14}$$

A.5 Expressions of transmission efficiency ζ and effective power supply P_{out}

$$\zeta = L_1 L_2 k^2 z_r \omega^2 \big/ D_1 \tag{A15}$$

$$D_1 = L_1 L_2 k^2 (R_{c2} + z_r) \omega^2$$
$$+ R_{c1}\left((R_{c2} + z_r)^2 + (z_i + L_2 \omega)^2 \right) \tag{A16}$$

$$P_{out} = 0.5\, V^2 L_1 L_2 k^2 z_r \omega^2 \big/ D_2 \tag{A17}$$
$$D_2 = 2 L_1 L_2 k^2 R_{c1} (R_{c2} + z_r) \omega^2$$
$$+ R_{c1}^2 \left((R_{c2} + z_r)^2 + (z_i + L_2 \omega)^2 \right)$$
$$+ L_1^2 \omega^2 \left((R_{c2} + z_r)^2 + \left(z_i - \left(-1 + k^2\right) L_2 \omega \right)^2 \right) \tag{A18}$$

A.6 Derivation of Eq.(19) and (20)

Eq.(A15) does not include z_i and the denominator Eq.(A16) becomes minimum when $z_i = -L\omega$. Therefore, z_i^e is determined independently of z_r^e as:

$$z_i^e = -L_2 \omega. \tag{A19}$$

The transmission efficiency ζ_{opti} at this z_i^e becomes:

$$\zeta_{opti} = \zeta \,|_{z_i = z_i^e} = \frac{z_r}{A z_r^2 + B z_r + C}. \tag{A20}$$

$$\left(\begin{array}{l} A = \dfrac{R_{c1}}{L_1 L_2 k^2 \omega^2} \\[2mm] B = 1 + \dfrac{2 R_{c1} R_{c2}}{L_1 L_2 k^2 \omega^2} \\[2mm] C = R_{c2} + \dfrac{R_{c1} R_{c2}^2}{L_1 L_2 k^2 \omega^2} \end{array} \right)$$

At the maximum point of above (A20), z_r satisfies $(d\,\zeta_{opti}/d\,z_r) = 0$. Therefore,

$$\frac{d\,\zeta_{opti}}{d\,z_r} = \frac{-A z_r^2 + C}{(A z_r^2 + B z_r + C)^2} = 0$$

$$z_r^e = \sqrt{C/A} = \sqrt{R_{c2}\left(R_{c2} + \frac{L_1 L_2 k^2 \omega^2}{R_{c1}} \right)}. \tag{A21}$$

This z_r^e gives the maximum ζ_m. Therefore,

$$\xi_m = \xi_{opti}\Big|_{z_r=z_r^e} = \xi\,|_{z_i=z_i^e,\,z_r=z_r^e}$$

$$= 1 - 2\,\frac{\sqrt{R_{c1}R_{c2}}\sqrt{R_{c1}R_{c2} + L_1L_2k^2\omega^2} - R_{c1}R_{c2}}{L_1L_2k^2\omega^2}.$$

(A22)

Eq.(19) is derived by assigning $R_{c1} = \sigma_1 L_1$, $R_{c2} = \sigma_2 L_2$ to the above (A22) and by some arrangement. Eq.(20) is derived from (A21) and (A19).

A.7 Derivation of Eq.(21) and (22)

The derivation processes of Eq.(21) and (22) are similar to that of Eq.(19) and (20). First, the denominator (A18) of P_{out} (A17) is arranged with coefficients U, V, and W that do not include z_i. Then,

$$D_2 = U + V\,(z_i + W)^2$$

(A23)

$$\left(\begin{array}{l} U = \dfrac{2\left(R_{c1}^2(R_{c2}+z_r) + L_1\left(L_2k^2R_{c1} + L_1(R_{c2}+z_r)\right)\omega^2\right)^2}{R_{c1}^2 + L_1^2\omega^2} \\[4mm] V = 2\left(R_{c1}^2 + L_1^2\omega^2\right) \\[4mm] W = L_2\,\omega\,\dfrac{R_{c1}^2 + L_1^2(1 - k^2)\omega^2}{R_{c1}^2 + L_1^2\omega^2} \end{array}\right)$$

is given. Therefore,

$$z_i^p = -W = -L_2\,\omega\,\frac{R_{c1}^2 + L_1^2(1 - k^2)\omega^2}{R_{c1}^2 + L_1^2\omega^2}$$

(A24)

is derived. The effective power supply P_{opti} at this z_i^p is arranged with X, Y, and Z that do not include z_r. Then,

$$P_{opti} = P_{out}\,|_{z_i=z_i^p} = \frac{z_r}{Xz_r^2 + Yz_r + Z}$$

(A25)

$$\left(\begin{array}{l} X = \dfrac{4R_{c1}^2R_{c2} + 4L_1\left(L_2k^2R_{c1} + L_1R_{c2}\right)\omega^2}{V^2L_1L_2k^2\omega^2} \\[4mm] Y = \dfrac{4R_{c1}^2R_{c2} + 4L_1\left(L_2k^2R_{c1} + L_1R_{c2}\right)\omega^2}{V^2L_1L_2k^2\omega^2} \\[4mm] Z = \dfrac{2\left(R_{c1}^2R_{c2} + L_1\left(L_2k^2R_{c1} + L_1R_{c2}\right)\omega^2\right)^2}{V^2L_1L_2k^2\omega^2\left(R_{c1}^2 + L_1^2\omega^2\right)} \end{array}\right)$$

is given. Therefore,

$$\frac{d\,P_{opti}}{d\,z_r} = \frac{-Xz_r^2 + Z}{(Xz_r^2 + Yz_r + Z)^2} = 0$$

$$z_r^p = \sqrt{Z/X} = \frac{R_{c1}^2 R_{c2} + L_1 \left(L_2 k^2 R_{c1} + L_1 R_{c2}\right)\omega^2}{R_{c1}^2 + L_1^2 \omega^2} \tag{A26}$$

is derived. This z_r^p gives the maximum P_m. Therefore,

$$P_m = P_{opti}\Big|_{z_r = z_r^p} = P_{out}\big|_{z_i = z_i^p,\, z_r = z_r^p}$$

$$= \frac{V^2 L_1 L_2 k^2 \omega^2}{8\left(R_{c1}^2 R_{c2} + L_1 \left(L_2 k^2 R_{c1} + L_1 R_{c2}\right)\omega^2\right)}. \tag{A27}$$

(21) is derived by assigning $R_{c1} = \sigma_1 L_1$, $R_{c2} = \sigma_2 L_2$ to the above (A27) and by some arrangement. (22) is derived from (A26) and (A24).

7. References

H. A. Haus (1984). *Waves and Fields in Optoelectronics*, Prentice-Hall, New Jersey

W. H. Hayt, Jr. (1989). *Engineering Electromagnetics*, McGraw-Hill

A. Karalis; J. D. Joannopoulos & M. Soljacic (2008). Efficient Wireless Non-Radiative Mid-Range Energy Transfer, *Annals of Physics*, Vol. 323, No. 1, pp. 34-48

M. K. Kazimierczuk (2009). *High-Frequency Magnetic Components*, John Wiley & Sons Ltd

A. Kurs; et al. (2007). Wireless Power Transfer via Strongly Coupled Magnetic Resonances, *Science Express*, Vol. 317, No. 5834, pp. 83-86

P. C. Maqnusson, et al. (2000). *Transmission Lines and Wave Propagation, Fourth Edition*, CRC Press

W. L. Stutzman (2011). *Antenna Theory and Design*, John Wiley & Sons Ltd

H. Sugiyama, Optimal Designs for Wireless Energy Link Based on Nonradiative Magnetic Field, *Proceedings of 13th IEEE Int. Symp. on Consumer Electronics*, Kyoto, Japan, May, 2009

H. Sugiyama, Autonomous Chain Network Formation by Multi-Robot Rescue System with Ad Hoc Networking, *Proceedings of IEEE Int. Workshop on Safety, Security, and Rescue Robotics*, Bremen, Germany, July 2010

H. Sugiyama, Optimal Designs for Wireless Energy Link with Symmetrical Coil Pair, *Proceedings of IEEE Int. Microwave Workshop Series on Innovative Wireless Transmission*, Kyoto, Japan, May 2011

7

Compact and Tunable Transmitter and Receiver for Magnetic Resonance Power Transmission to Mobile Objects

Takashi Komaru, Masayoshi Koizumi, Kimiya Komurasaki,
Takayuki Shibata and Kazuhiko Kano
DENSO CORPORATION and the University of Tokyo
Japan

1. Introduction

As electronic devices are becoming more mobile and ubiquitous, power cables are turning to the bottlenecks in the full-fledged utilization of electronics. While battery capacities are reaching their limits, wireless power transmission with magnetic resonance is expected to provide a breakthrough for this situation by enabling power feeding available anywhere and anytime.

This chapter studies the feasibility of magnetic resonance power transmission to mobile objects mainly focusing on the resonator quality factor and impedance matching control systems. Transmission efficiency reaches a reasonably high level when the transmitting and receiving resonators satisfy two conditions. The first is to have high quality factors. The second is to tune and match the impedance to the transmission distances. The second section explains the theoretical grounds for these conditions. The third section describes a developed wireless power transmission system prototype which was made compact and tunable to be applied to mobile objects. The later sections evaluate the quality factor and the impedance matching of the prototype.

2. Theoretical analysis of a magnetic resonance system

This section states the theoretical basics of wireless power transmission with magnetic resonance. The theory is developed by using the logic of electrical engineering without the coupled-mode theory. First, the base concept of a wireless power transmission system and its equivalent circuit model are introduced. Then the transmission efficiency is derived as a formula with the physical properties including the impedances and the mutual inductance. This formula is simplified by replacing the physical properties with the non-dimensional parameters including impedance ratio, quality factor and coupling coefficient. Analysis of this simplified formula derives the essential principle for efficient mid-range wireless power transmission.

2.1 Basic model

The model of a wireless power transmission system with magnetic resonance is expressed as per Fig. 1 (a) and (b). Two resonators of a series LCR (inductor, capacitor and resister) are inductively coupled to each other. Note that this model also applies to the basic system of wireless power transmission with electromagnetic induction. The currents flowing through the source and load are derived from Kirchhoff's second law as per (1).

$$\begin{bmatrix} V_{src} \\ 0 \end{bmatrix} = \begin{bmatrix} Z_S & j\omega M_{SD} \\ j\omega M_{SD} & Z_D + Z_{0ld} \end{bmatrix} \begin{bmatrix} I_{src} \\ I_{ld} \end{bmatrix}$$

$$\Rightarrow \begin{bmatrix} I_{src} \\ I_{ld} \end{bmatrix} = \frac{V_{src}}{Z_S(Z_D + Z_{0ld}) + (\omega M_{SD})^2} \begin{bmatrix} Z_D + Z_{0ld} \\ -j\omega M_{SD} \end{bmatrix} \tag{1}$$

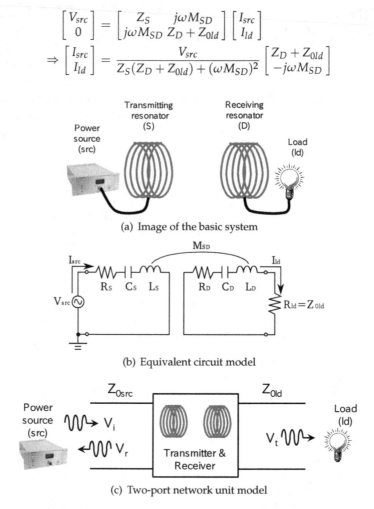

(a) Image of the basic system

(b) Equivalent circuit model

(c) Two-port network unit model

Fig. 1. Three types of concept figures of the basic wireless power transmission system model

Note that ω represents the angular frequency. Voltage V_{src} and currents I_{src}, I_{ld} are in phasor form and are complex properties. Impedance Z_S is defined as the impedance of the transmitting resonator $R_S + j(\omega L_S 1/\omega C_S)$. Impedance Z_D is also defined in the same way as Z_S. Z_{0src} represents the characteristic impedance of the transmission line in the power source device and the cable between the source and transmitting resonator. Impedance Z_{0ld} represents the characteristic impedance of the transmission line in the cable between the load and receiving resonator. These characteristic impedances are assumed to be a real number.

The load resistance R_{ld} is matched to Z_{0ld} so that the load absorbs all transmitted power and reflects no power to the receiving resonator.

Since the source frequency is often set to a high value of about 10 MHz, the concepts of incident wave, reflected wave and transmitted wave should be introduced (Pozer, 1998). As Fig. 1 (c) shows, the coupled resonators inserted between the source and the load can be considered as a two-port network unit. This unit receives incident waves from the source. At the same time, it produces reflected waves to the source and transmitted waves to the load. Voltage V_i, V_r and V_t represent the amplitude of each wave respectively. And they are expressed by the properties of the circuit model as per (2)

$$V_i = \frac{V_{src} + Z_{0src}I_{src}}{2}, V_r = \frac{V_{src} - Z_{0src}I_{src}}{2}, V_t = Z_{0ld}I_{ld} \tag{2}$$

The power of each wave is as per (3).

$$P_i = \frac{|V_i|^2}{2Z_{0src}}, P_r = \frac{|V_r|^2}{2Z_{0src}}, P_t = \frac{|V_t|^2}{2Z_{0ld}} \tag{3}$$

Then the transmission efficiency is defined as the ratio of transmitted wave power and incident wave power.

$$\eta = \frac{P_t}{P_i} = \frac{4Z_{0src}Z_{0ld}(\omega M_{SD})^2}{|(Z_S + Z_{0src})(Z_D + Z_{0ld}) + (\omega M_{SD})^2|^2} \tag{4}$$

The reflection efficiency is also defined as the ratio of reflected wave power and incident wave power.

$$\eta_r = \frac{P_r}{P_i} = \left| \frac{(Z_S - Z_{0src})(Z_D + Z_{0ld}) + (\omega M_{SD})^2}{(Z_S + Z_{0src})(Z_D + Z_{0ld}) + (\omega M_{SD})^2} \right|^2 \tag{5}$$

Equation (4) is the fundamental definition of efficiency. However, in some cases such as low frequency operations, the power source can reuse the reflected wave. Then the transmission efficiency becomes $\eta' = \eta/(1 - \eta_r)$, which is equal to the efficiency derived from a simple AC circuit calculation as per (6). Note that θ represents the phase difference between V_{src} and I_{src}.

$$\eta' = \frac{\eta}{1 - \eta_r} = \frac{Z_{0ld}|I_{ld}|^2}{|V_{src}I_{src}|\cos\theta} = \frac{Z_{0ld}|I_{ld}|^2}{R_S|I_{src}|^2 + (R_D + Z_{0ld})|I_{ld}|^2} \tag{6}$$

2.2 Analysis of transmission efficiency

The following non-dimensional parameters are useful when considering the transmission efficiency analytically.

- Coupling coefficient $k = M_{SD}/(L_S L_D)^{1/2}$
- Quality factor $Q_S = \omega_0 L_S/R_S$ and $Q_D = \omega_0 L_D/R_D$
- Impedance ratio $r_S = Z_{0src}/R_S$ and $r_D = Z_{0ld}/R_D$

Note that the resonant frequencies are assumed to be equal to each other ($(L_S C_S)^{-1/2} = (L_D C_D)^{-1/2} = \omega_0$) for simplicity in this discussion. The frequency band is considered narrow enough ($|\omega - \omega_0| \ll \omega_0$). Then (4) is expressed as (7) by using the approximations $\omega/\omega_0 - \omega_0/\omega \approx 2(\omega - \omega_0)/\omega_0$ and $\omega \approx \omega_0$.

$$\eta \approx \frac{4k^2 \frac{r_S}{Q_S} \frac{r_D}{Q_D}}{\left[k^2 - 4\left(\frac{\omega - \omega_0}{\omega_0}\right)^2 + \frac{1+r_S}{Q_S}\frac{1+r_D}{Q_D} \right]^2 + 4\left(\frac{\omega - \omega_0}{\omega_0}\right)^2 \left(\frac{1+r_S}{Q_S} + \frac{1+r_D}{Q_D}\right)^2} \tag{7}$$

Efficiency η' is also expressed by the non-dimensional parameters.

$$\eta' = \frac{k^2 Q_S Q_D r_D}{(1 + r_D)(1 + r_D + k^2 Q_S Q_D) + 4[(\omega - \omega_0)/\omega_0]^2 Q_D^2} \tag{8}$$

In the $\omega - r_S - r_D$ domain, η has its maximum value (9) at (10).

$$\eta_{max} = \frac{k^2 Q_S Q_D}{\left(1 + \sqrt{1 + k^2 Q_S Q_D}\right)^2} \tag{9}$$

$$\omega = \omega_0, \quad r_S = r_D = \overline{1 + k^2 Q_S Q_D} \tag{10}$$

The physical meaning of (10) is the impedance matching condition. In the case $r_S = r_D$, η in the $\omega - r_S - r_D$ domain becomes visible as the surface in Fig. 2, which clearly shows the existence of a maximum point. It is not easy to prove directly from (7) that (9) is exactly the maximum. However, there is a simple and strict proof via analysis at (8).

$$\eta' \leq \frac{k^2 Q_S Q_D r_D}{(1 + r_D)(1 + r_D + k^2 Q_S Q_D)} \tag{11}$$

$$\leq \frac{k^2 Q_S Q_D}{(1 + \sqrt{1 + k^2 Q_S Q_D})^2} \equiv \eta'_{max} \tag{12}$$

The equality in (11) and (12) is approved when $\omega = \omega_0$ and $r_D = (1 + k^2 Q_S Q_D)^{1/2}$ respectively. These directly mean the maximum of η' is $\eta'_{max} = \eta_{max}$ at $\omega = \omega_0$ and $r_D = (1 + k^2 Q_S Q_D)^{1/2}$. Besides $\eta \leq \eta'$ because $\eta' \leq \eta/(1 - \eta_r)$ and $0 \leq \eta_r \leq 1$. Thus the maximum value of (7) is proved to be (9) as per (13).

$$\eta \leq \eta' \leq \eta'_{max} = \eta_{max} \quad \Rightarrow \quad \eta \leq \eta_{max} \tag{13}$$

Equation formula (9) depends only on one term $k(Q_S Q_D)^{1/2}$ and draws a simple curve as shown in Fig. 3.
Here the principles for efficient power transmission are derived without referring to the coupled-mode theory.

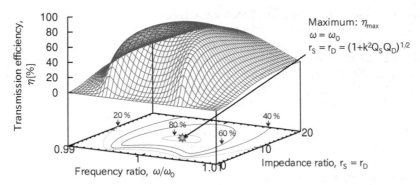

Fig. 2. Three dimensional plot and contour of the transmission efficiency in $\omega - r$ domain with $k = 0.01$ and $Q_S = Q_D = 1000$.

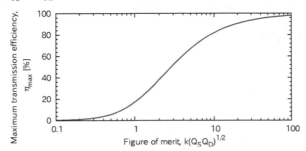

Fig. 3. Maximum transmission efficiency plotted along figure-of-merit

- $k(Q_S Q_D)^{1/2}$ is defined as figure-of-merit (*fom*).

- Impedance ratio should be matched to $(1 + fom^2)^{1/2}$.

- Figure-of-merit in 1 or higher order enables efficient wireless power transmission with over 20% efficiency. (The concept of a strong coupling regime.)

They completely correspond to the characteristics of mid-range transmission argued in the original theory of magnetic resonance (Karalis et al., 2008). In fact, $k(Q_S Q_D)^{1/2}$ can be derived from the original expression of figure-of-merit $\kappa/(\Gamma_S \Gamma_D)^{1/2}$ (Kurs et al., 2007). In the field of electrical engineering, coupling coefficient k and the quality factor Q are more intuitive parameters than the reciprocally-time-dimensional coupling coefficient κ and decay constant Γ. Thus the redefinition of figure-of-merit helps the understanding of mid-range transmission in comparison with conventional close-range electromagnetic inductance.

In close-range electromagnetic inductance, k is over 0.8 in most cases and rarely less than 0.2. Therefore, it is relatively easy to achieve high transmission efficiency of over 80% (Ayano et al., 2003; Ehara et al., 2007). Meanwhile in the mid-range, the transmission distance is equal to or greater than the size of resonators and k becomes a very low value of 10^2 or lower order. Here, the principle of the strong coupling regime concept leads to the point of importance: it is a high quality factor of 10^2 or greater order that enables efficient transmission even with very low k in mid-range.

In addition to high quality factor, impedance matching is needed for efficient transmission in mid-range. Mid-range wireless power transmission is expected to bring much more flexibility in receiver and transmitter positioning than close range wireless power transmission. And this means the coupling coefficient between transmitting and receiving resonators can take various values. Moreover in mobile applications, receivers and transmitters will be equipped on moving objects and have to exchange power across ever-changing transmission distance and coupling coefficients. As stated before, the impedance ratio r_S and r_D must be matched to $(1+fom^2)^{1/2} = (1+k^2 Q_S Q_D)^{1/2}$ in order to obtain the highest efficiency. Fig. 4 shows how transmission efficiency changes with different coupling coefficients. When the receiver comes to the closer (a) or the farther (c) position than the original position (b), coupling coefficient becomes larger or smaller and the maximum point in w-r domain changes to the higher-r or the lower-r point. Thus we must consider a system that controls the impedance ratio according to the variable transmission distance.

There are three schemes considered for the impedance matching control: two-sided, one-sided and no control. In the two-sided control scheme, both the transmitter and the receiver take the optimum impedance ratios. In the one-sided control scheme, either the transmitter or the receiver takes the optimum one and the other takes a fixed one. In the no control scheme, both the transmitter and the receiver take fixed ones. The theoretical efficiencies with these control schemes are expressed as follows:

$$\eta_2 = \frac{k^2 Q_S Q_D}{(1+\sqrt{1+k^2 Q_S Q_D})^2} \tag{14}$$

$$\eta_1 = \frac{k^2 Q_S Q_D r_D}{(1+r_D)(1+r_D+k^2 Q_S Q_D)} \tag{15}$$

$$\eta_0 = \frac{4k^2 Q_S Q_D r_S r_D}{[(1+r_S)(1+r_D)+k^2 Q_S Q_D]^2} \tag{16}$$

Note that η_2, η_1 and η_0 represent the efficiency with two-sided, one-sided and no control scheme respectively. Equation 15 assumes that the transmitter has the optimized impedance ratio and the receiver has the fixed one. When the transmitter has the fixed one and the receiver has the optimized one in reverse, r_D and r_S switch positions with each other. The no control efficiency η_0 is derived from η in (7) by substituting w with w_0. The one-sided control efficiency η_1 is derived from the partial differential analysis of η_0 as follows.

$$\frac{\partial}{\partial r_S} \log \eta_{nctl} = 0 \Rightarrow k^2 Q_S Q_D + (1-r_S)(1+r_D) = 0 \tag{17}$$

Efficiency η_1 is equal to η' in (8) with $w = w_0$. This means the reuse of the reflected power by the power source is equal to the optimization of the transmitter impedance ratio. The two-sided control efficiency η_2 is exactly the same as η_{max}.

The characteristics of the transmission efficiencies for the two-sided, one-sided and no control are visualized by Fig. 5. Generally, two-sided control always takes the maximum efficiency, while it will take more hardware and software cost for the control system. No control will only need a simple and low-cost system in exchange for efficiency at various transmission distances. One-sided control has the middle characteristics.

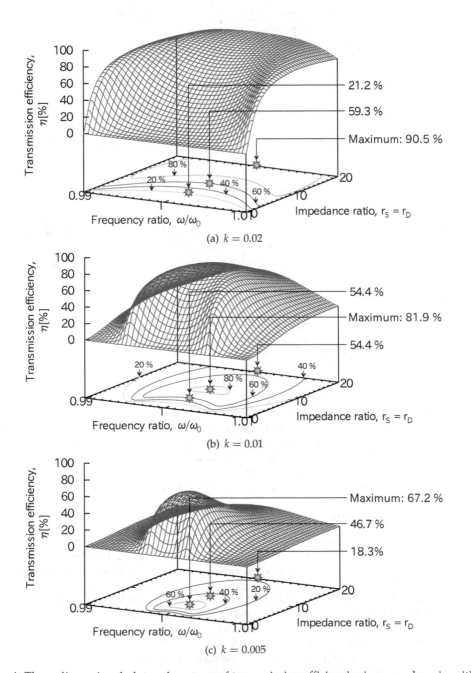

(a) $k = 0.02$

(b) $k = 0.01$

(c) $k = 0.005$

Fig. 4. Three dimensional plot and contour of transmission efficiencies in $\omega - r$ domain with various coupling coefficients k and quality factors $Q_S = Q_D = 1000$

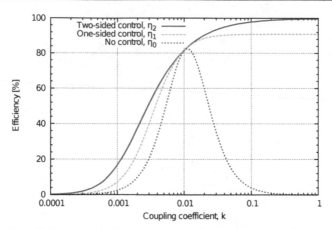

Fig. 5. Transmission efficiency with the three types of impedance matching control schemes. The quality factors are $Q_S = Q_D = 1000$. The impedance ratios are fixed as $r_S = [1 + (10^{-2})^2 Q_S Q_D]^{1/2}$ in the one-sided control and $r_S = r_D = [1 + (10^{-2})^2 Q_S Q_D]^{1/2}$ in the no control.

3. Specifications of the transmitter and the receiver

Efficient mid-range wireless power transmission with magnetic resonance needs a high quality factor resonator and impedance matching system. Mobile object applications need compact and tunable transmitters and receivers. We developed an experimental transmitter and receiver system that satisfy those conditions as shown in Fig. 6. The system consists of a resonator and a pickup loop. The resonator is a copper wire loop with a lumped mica capacitor. The pick up loop is also a copper wire loop with a lumped mica capacitor. The combination of the resonator and the pickup loop function as an inductive transformer, which virtually transforms the characteristic impedance of the source and the load. Both the resonator and the wire loop have only a single turn and they are placed in parallel and very close to each other. This structure makes the transmitter and receiver axially compact. Note that the resonant frequencies of the resonator and the pickup loop are designed to be equal to each other. The following two sections state the performance analyses of the system.

4. Resonator quality factor

Resonators may take any shape and any materials. A popular type of resonator is a coil with multiple turns. To make it compact, the turn pitch has to be as small as possible. But with such a dense coil structure, it is hard to have high quality factor because of the electricity loss due to the insulating cover on the wire (Komaru et al., 2010). The other widely considered type is a coil with a single turn and a lumped capacitor. In this study, we call this type "loop with capacitor" and analyze it as a primary model of the resonator.

4.1 Theoretical calculation
An exact calculation of quality factors is indispensable to design and evaluate the wireless power transmission system with magnetic resonance. As stated in the previous section, a resonator is equivalent to a series LCR element and quality factor Q is generally derived as

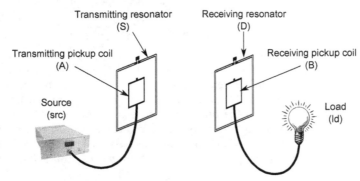

(a) Overall view of the power transmission system

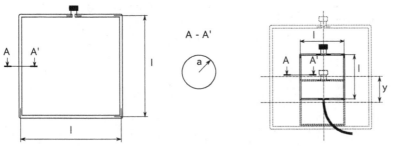

(b) The resonator with side length $\ell \times \ell = 198$ mm $\times$ 198 mm, wire radius $a = 1.5$ mm and capacitor capacitance: $C = 200$ pF.

(c) The pickup loop with side length $\ell \times \ell = 92$ mm $\times$ 92 mm, wire radius $a = 1.0$ mm, capacitor capacitance $C = 470$ pF and sliding position $y = 0$ - 48 mm.

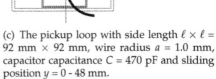

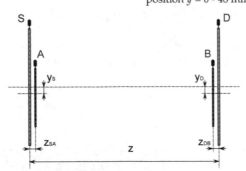

(d) Alignment of the resonators and pickup loops with the pickup loop slide positions $y_S, y_D = 0$ - 48 mm and the spaces between resonator and pickup loop $Z_{SA} = Z_{DB} = 2.0$ mm.

Fig. 6. Transmitter and receiver specification

the ratio of resonant frequency $f_0 = \omega_0/2\pi = 1/2\pi(LC)^{1/2}$, self inductance L and resistance R of a resonator.

$$Q = \frac{\omega_0 L}{R} \tag{18}$$

Of these properties, resistance is the most difficult to calculate regarding its realizable value. The predictions in most conventional research only take account of two types of resistance: radiation and ohmic resistance. This chapter also calculates capacitor resistance to predict the quality factor precisely. Then the properties to be calculated are resonant frequency, self inductance, radiation resistance ohmic resistance and capacitor resistance. The first four properties are calculated by the theoretical equations derived from the electromagnetism. The other capacitor resistance is estimated by using the measured specification data of a capacitor element.

The total length of the resonator wire is about 0.8 m and is far smaller than the wavelength of about 30 m in the supposed frequency band. Thus the current distribution on the wire is approximated to be uniform hereinafter.

For a rectangle loop with a side length ℓ_x, ℓ_y and a wire radius a, the self inductance is expressed as (19) (Grover, 1946).

$$
\begin{aligned}
L = \frac{\mu_0}{\pi} & \left[\ell_x \ln\left(\frac{2\ell_x}{a}\right) + \ell_x \ln\left(\frac{2\ell_y}{a}\right) + 2 \cdot \sqrt{\ell_x^2 + \ell_y^2} \right. \\
& \left. -\ell_x \sinh^{-1}\left(\frac{\ell_x}{\ell_y}\right) - \ell_y \sinh^{-1}\left(\frac{\ell_y}{\ell_x}\right) - 1.75(\ell_x + \ell_y) \right]
\end{aligned}
\tag{19}
$$

Substituting ℓ_x and ℓ_y with ℓ leads to

$$L = \frac{2\mu_0 \ell}{\pi}\left[\ln\left(\frac{2\ell}{a}\right) - 1.2172 \right] \tag{20}$$

The uniform current distribution means zero electrical charge density at every point on the wire. Therefore the resonator capacitance originates just from the lumped capacitor. Then the resonant frequency is

$$\omega_0 = \frac{1}{\sqrt{LC}} \tag{21}$$

The resonator is approximated as a small current loop, which is theoretically equal to a small magnetic dipole as in Fig. 7, because of the loop size being far smaller than the wavelength. The radiation resistance is derived from consideration of this small magnetic dipole model. A small electrical dipole with current I and length dz radiates power expressed as

$$\frac{\pi |I|^2}{3}\sqrt{\frac{\mu_0}{\epsilon_0}}\left(\frac{dz'}{\lambda_0}\right)^2 \tag{22}$$

Note that λ_0 is the vacuum wavelength $c_0/f_0 = 2\pi c_0/\omega_0$. According to the symmetry of Maxwell's equations, ϵ_0, μ_0 and I are the dual of μ_0, ϵ_0 and I_m respectively. I_m represents

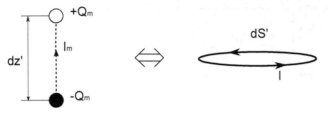

Fig. 7. Magnetic dipole and electrical current loop

the magnetic current (Johnk, 1975). Then the radiated power from a small magnetic dipole is expressed as

$$P_{rad} = \frac{\pi |I_m|^2}{3} \sqrt{\frac{\epsilon_0}{\mu_0}} \left(\frac{dz'}{\lambda_0}\right)^2 \qquad (23)$$

From the definition of the magnetic dipole moment of both the magnetic dipole and the electrical current loop

$$Q_m dz' = \mu_0 I dS' \qquad (24)$$

where dS' is the loop area corresponding to ℓ^2 and Q_m is the magnetic charge with the relation

$$I_m = \frac{\partial Q_m}{\partial t} = j\omega Q_m \qquad (25)$$

The radiated power is then transformed to

$$P_{rad} = \frac{\mu_0 \omega^4 \ell^4}{12\pi c_0^3} |I|^2 \qquad (26)$$

which should be equal to

$$P_{rad} = \frac{1}{2} R_{rad} |I|^2 \qquad (27)$$

Hence the radiation resistance is

$$R_{rad} = \frac{\mu_0 \omega^4 \ell^4}{6\pi c_0^3} \qquad (28)$$

The basic expression of the ohmic resistance for a wire with conductivity σ, sectional area S and length $d\ell$ is

$$R_{ohm} = \frac{d\ell}{\sigma S} \qquad (29)$$

When the frequency is high, current density decays exponentially along the depth from the surface to the center of the wire. The skin depth δ_s is defined as the inverse of the decay constant (Pozer, 1998) as

$$\delta_s = \sqrt{\frac{2}{\omega \mu \sigma}} \qquad (30)$$

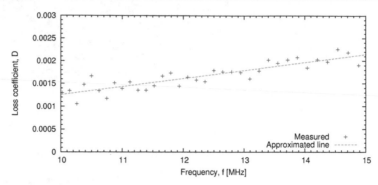

Fig. 8. Characteristics of the capacitor loss coefficient

and wire sectional area S in (30) is substituted with $S = 2\pi a \delta_s$

$$R_{ohm} = \frac{d\ell}{2\pi a}\sqrt{\frac{\mu\omega}{2\sigma}} \tag{31}$$

the total wire length of the resonator is 4ℓ and the ohmic resistance finally becomes

$$R_{ohm} = \frac{\ell}{\pi a}\sqrt{\frac{2\mu\omega}{\sigma}} \tag{32}$$

A real capacitor is not a pure capacitive element but a complex component with various impedance elements. In the frequency band of wireless power transmission, the dominant characteristics are the original capacitance and the equivalent series resistance (ESR). This study calls ESR "capacitor resistance" and analyzes its value. The characteristics of the actual capacitor were measured by the LCR meter. The meter gave the loss coefficient data as in Fig. 8. Note that the approximated line is derived by using the least-squares method. The loss coefficient of the capacitor is defined as

$$D(\omega) = -\frac{R(\omega)}{X(\omega)} = \omega CR \tag{33}$$

Note that X and R here represent the reactance and the resistance of the capacitor respectively. The capacitor resistance is then expressed as

$$R_{cap} = \frac{D(\omega)}{\omega C} \tag{34}$$

4.2 Experimental measurement

The comparison with the calculated and the measured quality factors are shown in Fig. 9 in the form of the loss coefficient. Note that the loss coefficient is the inverse of the quality factor. The calculated loss coefficient is classified into the three categories corresponding to the three resistances introduced in the previous subsection. The radiation loss is too small to be visible in this graph. The predicted resonant frequencies and the quality factors are well agreed with the measured parameters, verifying the calculation method stated in this section. The important point is the quality factor of the loop with the capacitor type resonator was

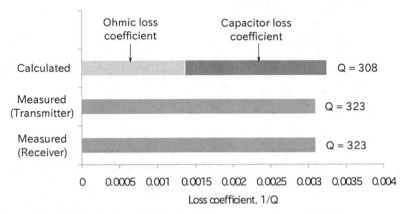

Fig. 9. Comparison between the calculated and the measured quality factors

higher than that of the dense coil type resonator at about $Q = 200$ as measured in our previous study. And the capacitor loss was as large as the ohmic loss. This means that the calculation of the capacitor loss was indispensable. Note that the measured resonant frequencies of the transmitting and the receiving resonator were 13.43 and 13.46 MHz. They also agreed with the calculated value of 13.54 MHz.

5. Impedance matching control system

Impedance matching means the transformation of the source and the load impedance. The impedance matching system has the same purpose as the antenna tuners. There are several ways to implement impedance matching: some use a set of variable capacitors and inductors and some others use transistors for example. The main requirement of the impedance matching system for efficient transmission is low electrical loss. And for mobile applications, it is also required to be compact and have wide range of impedance transformation. This chapter considers a sliding pickup loop system as a simple and compact transmitter and receiver prototype.

5.1 Basics of impedance matching with pickup loop

Impedance transformation is explained as follows. The coupling of the resonator and the pickup loop is equivalent to the circuit model shown in Fig. 10. Though the indices here are for the receiving resonator and the pickup loop, the same result is obtained for the transmitting resonator and the pickup loop. And the following mutual inductances are assumed to be negligible: M_{AB} between the transmitting and the receiving pickup loop, M_{SB} between the transmitting resonator and the receiving pickup loop and M_{DA} between the receiving resonator and the transmitting pickup loop. From Kirchhoff's second law for the circuit, the currents flowing in the resonator and the pickup loop I_D, I_{ld} is expressed by the voltage in the resonator V which the current in the pickup loop inducted.

$$\begin{bmatrix} V \\ 0 \end{bmatrix} = \begin{bmatrix} 0 & j\omega M_{DB} \\ j\omega M_{DB} & Z_B + Z_{0ld} \end{bmatrix} \begin{bmatrix} I_D \\ I_{ld} \end{bmatrix}$$

$$\Rightarrow \begin{bmatrix} I_D \\ I_{ld} \end{bmatrix} = \frac{V}{(\omega M_{DB})^2} \begin{bmatrix} Z_B + Z_{0ld} \\ -j\omega M_{DB} \end{bmatrix} \tag{35}$$

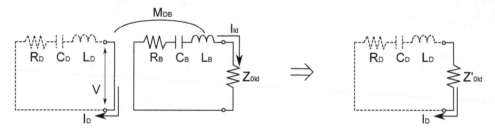

Fig. 10. Circuit model of the impedance transformation with pickup loop

Then the pickup loop and the load that are inductively coupled to the resonator equal to the impedance Z_{Old} directly connected to the resonator.

$$Z'_{Old} = \frac{V}{I_D} = \frac{(\omega M_{DB})^2}{Z_B + Z_{Old}} \tag{36}$$

Note that Z_B represents the impedance of the pickup loop $Z_B = R_B + i(\omega L_B 1/\omega C_B)$. Let the resonant frequency of the pickup loop equal the resonant frequency of the resonator and the source frequency so that the pickup loop reactance is zero $Z_B = R_B$. Now the impedance ratio becomes

$$r_D = \frac{Z'_{Old}}{R_D} = \frac{V}{R_D I_D} = \frac{(\omega M_{DB})^2}{R_D(R_B + Z_{Old})} \tag{37}$$

This is the function of the mutual inductance between the resonator and the pickup loop. This means it is possible to control the impedance ratio by changing the coupling condition, including the relative position, of the resonator and the pickup loop.

To design a transmitter or a receiver with this pickup loop impedance matching system and acquire the desired impedance ratios, the following properties have to be calculated: the resistance R_B, the self inductance L_B, the capacitance C_B of the pickup loop and the mutual inductance between the resonator and the pickup loop M_{DB}. For the pickup loop with a square loop wire and a lumped capacitor, the first three properties are calculated by the same method introduced in previous section. The last property is calculated from electromagnetic theory as explained in the following subsection.

5.2 Mutual inductance between square loops

For coupling of square loop wires, the theoretical formula of mutual inductance is derived from the Neumann formula (38), which expresses the mutual inductance of coupled circuits s_1 and s_2 in free space, as illustrated in Fig. 11 (a).

$$\frac{\mu_0}{4\pi} \int_{s_1} \int_{s_2} \frac{d\vec{s_1} \cdot d\vec{s_2}}{|\vec{s_1} - \vec{s_2}|} \tag{38}$$

Then the mutual inductance between two parallel wires of finite line, as shown in Fig. 11 (b), is derived by using the Neumann formula as

$$M_\parallel(\ell_1, \ell_2, \xi, \zeta) = m_\parallel(\ell_1, \ell_2, \xi, \zeta) + m_\parallel(\ell_1, \ell_2, -\xi, \zeta)$$
$$- m_\parallel(\ell_1, -\ell_2, \xi, \zeta) - m_\parallel(\ell_1, -\ell_2, -\xi, \zeta) \tag{39}$$

The abbreviation $m_\parallel$ is defined as

$$m_\parallel(\ell_1, \ell_2, \xi, \zeta) \equiv \frac{\mu_0}{8\pi} \left[(\ell_1 + \ell_2 + 2\xi) \sinh^{-1} \frac{\ell_1 + \ell_2 + 2\xi}{2\zeta} - \sqrt{(\ell_1 + \ell_2 + 2\xi)^2 + (2\zeta)^2} \right] \tag{40}$$

Now consider the couple of parallel square loop illustrated in Fig. 11 (c). The couple consists of 8 lines: line a to d and A to D. According to the Neumann formula, the mutual inductance of the couple is expressed by the summation of the mutual inductances among them.

$$M = \sum_{i=A,B,C,D} \sum_{j=a,b,c,d} M_{ij} \tag{41}$$

M_{ij} represents the mutual inductance between the line i either of the line A to D and the line j either of the line a to d. The mutual inductances between the perpendicular lines such as M_{Ab} are all zero and those between the parallel lines including M_{Aa} are expressed by using (39). Hence the mutual inductance of the square loops is derived as

$$M = M_\parallel \left[\ell_1, \ell_2, x, \sqrt{(\bar{\ell}_- + y)^2 + z^2} \right] - M_\parallel \left[\ell_1, \ell_2, x, \sqrt{(\bar{\ell}_+ + y)^2 + z^2} \right]$$
$$+ M_\parallel \left[\ell_1, \ell_2, y, \sqrt{(\bar{\ell}_- - x)^2 + z^2} \right] - M_\parallel \left[\ell_1, \ell_2, y, \sqrt{(\bar{\ell}_+ - x)^2 + z^2} \right]$$
$$+ M_\parallel \left[\ell_1, \ell_2, -x, \sqrt{(\bar{\ell}_- - y)^2 + z^2} \right] - M_\parallel \left[\ell_1, \ell_2, -x, \sqrt{(\bar{\ell}_+ - y)^2 + z^2} \right]$$
$$+ M_\parallel \left[\ell_1, \ell_2, -y, \sqrt{(\bar{\ell}_- + x)^2 + z^2} \right] - M_\parallel \left[\ell_1, \ell_2, -y, \sqrt{(\bar{\ell}_+ + x)^2 + z^2} \right] \tag{42}$$

Note that the abbreviation $\bar{\ell}_\pm$ is defined as

$$\bar{\ell}_\pm \equiv \frac{\ell_1 \pm \ell_2}{2} \tag{43}$$

5.3 Experimental measurements

The measured wireless power transmission efficiencies with the three types of impedance ratio control are shown in Fig. 12. The theoretical curves are derived by using the measured coupling coefficient, quality factors and the theoretical formulas (14 - 16). Note that the frequency of the power source is 13.44 MHz, which is the average of the resonant frequency of the transmitting and receiving resonator.

The measurement results of the pickup loop slide position are shown in Fig. 13. The theoretical, which means optimum, curve of the impedance ratio is derived by using the measured coupling coefficient, quality factors and the theoretical conditions of the impedance

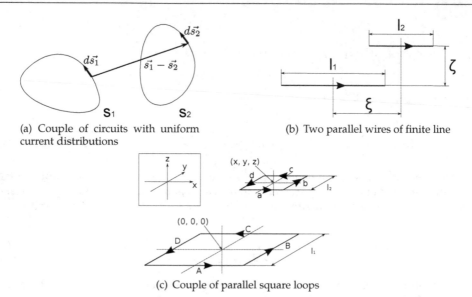

(a) Couple of circuits with uniform current distributions

(b) Two parallel wires of finite line

(c) Couple of parallel square loops

Fig. 11. Geometries applied for the Neumann formula to derive the mutual inductances

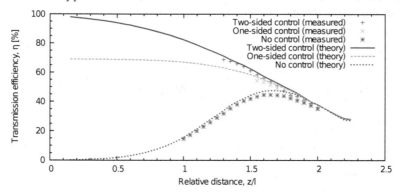

Fig. 12. Transmission efficiency with the three types of impedance matching control schemes. The relative distance is the ratio of a transmission distance z = 200 - 400 mm and a side length ℓ = 198 mm of the transmitters and the receiver.

matching. The conditions are stated as (10) for the two-sided control and as (17) for the one-sided control. The measured value of the impedance ratio is estimated by using the measured pickup loop slide position and the theoretical formula of the impedance transformation (37) with the calculated properties of the resonator and the pickup loop.

Each measured value well agreed with each theoretical value. Especially in the range of z/ℓ = 1.3 - 2.0 for the two-sided control, the wireless power transmission efficiency was as high as 30 - 70 % even though the coupling coefficient is lower than 0.02. Hence the effectiveness of wireless power transmission with magnetic resonance was verified in the mid-range. The optimum pickup loop slide positions became smaller when the transmission distances got larger. In the result, the corresponding impedance ratios became smaller as the

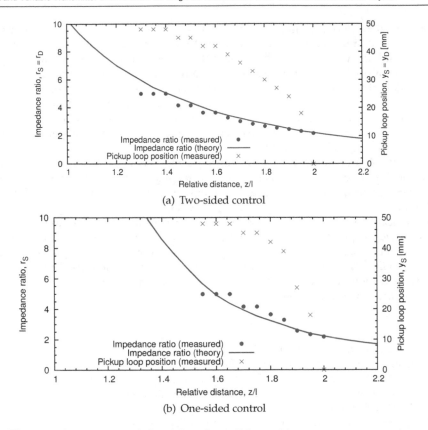

(a) Two-sided control

(b) One-sided control

Fig. 13. The impedance ratio and the pickup loop slide position

theory predicts. This is because the magnitude of the magnetic field inside the square loop is larger near the border and smaller near the center. The controllable transmission range is about $z/\ell = 1.3$ - 2.0 with the two-sided control and $z/\ell = 1.5$ - 2.0 with the one-sided control. The three efficiencies were equal at relative distance $z/\ell = 2.0$. When the transmission distance became shorter, efficiency with the no control rapidly dropped. On the other hand, efficiency with the one-sided control went up a little and soon became almost constant. These characteristics suggest a wireless power transmission system with magnetic resonance will have to carefully choose a proper impedance matching control scheme according to the moving range of mobile receivers in practical use. In narrow transmission distance range, no control would be enough. But in wide transmission distance range, one-sided control is needed at least. If the application needs much higher efficiency at shorter distances, the two-sided control should be implemented.

6. Conclusions

This chapter studied the feasibility of wireless power transmission with magnetic resonance to mobile objects.

The theory of magnetic resonance was analyzed not with the coupled-mode theory but with the electrical engineering theory to emphasize the essential elements of magnetic resonance:

high quality factor resonator and impedance matching system. The theory also derives the three schemes of impedance matching control. Two-sided control produces the maximum efficiency while it would require higher hardware and software costs. No control will only need a simple and low-cost system in exchange for efficiency at various transmission distances. One-sided control has middle characteristics between the first two.

A transmitter and receiver system prototype was developed to verify the theory and to discuss the realizable performance of a compactly-implemented resonator and impedance matching system. The resonator was a loop with capacitor type resonator and the impedance matching system was a sliding pickup loop system.

Evaluation of the resonator quality factor showed the loop with capacitor type resonator had a higher quality factor than the other compactly-shaped dense coil type resonator. And it was proven that the quality factor depends not only on radiation and ohmic loss but also on capacitor loss.

The theoretical analysis of the sliding pickup loop system and the power transmission experiment were explained. It showed that the couple of the pickup loop and the resonator functions as an inductive transformer and the sliding position of the pickup loop controls the impedance ratio. The power transmission experiment also verified the theory of wireless power transmission with magnetic resonance and the theoretical characteristics of tree impedance matching control schemes.

7. References

A. Karalis, J. D. Joannopoulos & M. Soljačić (2008). Efficient wireless non-radiative mid-range energy transfer, *Annals of Physics* Volume 323, Issue 1: 34–48.

A. Kurs, A. Karalis, R. Moffatt, J. D. Joannopoulos, P. Fisher & M. Soljačić (2007). Wireless Power Transfer via Strongly Coupled Magnetic Resonances, *Science Magazine* Volume 317(No. 5834): 83–86.

Carl T.A. Johnk (1975). *Engineering Electromagnetic Fields and Waves*, Wiley, New York

D. M. Pozer (1998). *Microwave Engineering, 2nd ed.*, Wiley, NY

Frederick Warren Grover (1946). *Inductance Calculations: Working Formulas and Tables*, Van Nostrand, New York

H. Ayano, H. Nagase & H. Inaba (2003). High Efficient Contactless Electrical Energy Transmission System, *IEEJ Tran. IA* Volume 123(No. 3): 263–270.

T. Komaru, M. Koizumi, K. Komurasaki, T. Shibata & K. Kano (2010). Parametric Evaluation of Mid-range Wireless Power Transmission with Magnetic Resonance, *Proceedings of the IEEE-ICIT 2010 International Conference on Industrial Technology*, pp. 789–792.

N. Ehara, Y. Nagatsuka, Y. Kaneko, S. Abe, T. Yasuda & K. Ida (2007). Compact and Rectangular Transformer of Contactless Power Transfer System for Electric Vehicle, *The Papers of Technical Meeting on Vehicle Technology*, pp. 7–12.

8

Realizing Efficient Wireless Power Transfer in the Near-Field Region Using Electrically Small Antennas

Ick-Jae Yoon and Hao Ling
Dept. of Electrical and Computer Engineering,
The University of Texas at Austin,
USA

1. Introduction

In the early 1900's, Tesla carried out his experiment on power transmission over long distances by radio waves (Tesla, 1914). He built a giant coil (200-ft mast and 3-ft-diameter copper ball positioned at the top) resonating at 150 kHz and fed it with 300 kW of low frequency power. However, there is no clear record of how much of this power was radiated into space and whether any significant amount of it was collected at a distant point. Over the years, wireless power delivery systems have been conceived, tried and tested by many (Brown, 1984; Glaser, 1968; McSpadden et al., 1996; Shinohara & Matsumoto, 1998; Strassner & Chang, 2002; Mickle, et al., 2006; Conner, 2007). For very short ranges, the inductive coupling mechanism is commonly exploited. This is best exemplified by non-contact chargers and radio frequency identification (RFID) devices operating at 13 MHz (Finkenzeller, 2003). Such systems are limited to ranges that are less than the device size itself. For long distance wireless power delivery, directed radiation is required, which dictates the use of large aperture antennas. This line of thinking is best exemplified by NASA's effort to collect solar power on a single satellite station in space and relay the collected power via microwaves to power other satellites in orbit (Glaser, 1968). In addition to needing large antennas, this scheme requires uninterrupted line-of-sight propagation and a potentially complicated tracking system for mobile receivers.

In 2007, MIT physicist Soljačić and his group demonstrated the feasibility of efficient non-radiative wireless power transfer using two resonant loop antennas (Kurs et al., 2007). Since then, there has been much interest from the electromagnetics community to more closely study this phenomenon (Kim & Ling, 2007; Jing & Wang, 2008; Kim & Ling, 2008; Pan et al., 2009; Thomas et al., 2010; Jung and Lee, 2010; Cannon et al., 2009; Kurs et al., 2010; Casanova et al., 2009). It was found that when two antennas are very closely spaced, they are locked in a coupled mode resonance phenomenon. In this coupled mode region, the two antennas see each other's presence strongly, and very high power transfer efficiency (PTE) can be attained (see Fig. 1). The term magnetic resonance coupling is often used to describe this phenomenon, although such coupled mode resonance can exist in antenna systems dominated by either magnetic or electric coupling, as shown in (Kim & Ling, 2007). It was also found that to maximize the power transfer the antennas need to have low radiation loss so that power is not lost to the radiation process. However, such a coupled mode region is very short when measured in terms of wavelength and attempts to design antennas to extend the distance over which such coupled mode phenomenon can be maintained have proven to be rather difficult (Kim & Ling, 2008). If the range can be extended, it could potentially benefit a number of applications including RFID (Fotopoulou & Flynn, 2007; Fotopoulou & Flynn, 2011; Sample & Smith, 2009; Bolomey et al., 2010; Yates et al., 2004), biomedical implants (Poon et al., 2010; RamRakhyani et al., 2011; Smith et al., 2007; Fotopoulou & Flynn, 2006), and electric car charging (Eom & Arai, 2009; Imura et al., 2009).

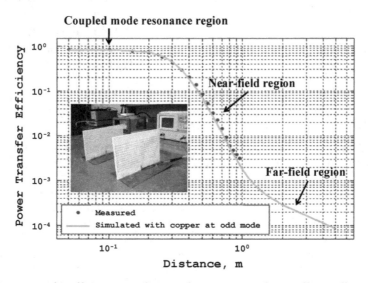

Fig. 1. Power transfer efficiency vs. distance between two electrically small meander antennas operating at 43MHz (Kim & Ling, 2007).

When the distance between the two antennas increases, the coupled mode resonance phenomenon disappears, and the antennas behave like traditional transmitting and receiving antennas in the near field (see Fig. 1). Beyond the coupled mode region, the PTE decreases rapidly as a function of distance. Nevertheless, sufficiently high PTE values might still be maintained for such coupling to be useful for power transfer. The physics beyond the coupled mode region is significantly easier to interpret. There have been works on the coupling properties of antennas in the near-field region, though the interest was not wireless power transfer. In the 1970's, theoretical and practical aspects of the spherical near-field antenna measurement were extensively studied at the Technical University of Denmark (Hansen, 1988). Yaghjian formulated a coupling equation between a transmitting and receiving antenna pair by the spherical wave expansion (Yaghjian, 1982). To more accurately estimate the coupling in the near-filed region, heuristic corrections to the far-field Friis formula were also attempted by adding higher order distance terms (Schantz, 2005) or by adding a gain reduction factor (Kim et al., 2010). Recently, a theoretical PTE bound for wireless power transfer in the near-field region was presented under the optimal load condition by Lee and Nam (Lee & Nam, 2010). They applied the spherical mode theory to study the coupling between two electrically small antennas in this region. It was found that a 40% PTE value can be theoretically achieved at a distance of 0.26λ in the co-linear configuration of electric dipole antennas, while the same PTE value was obtained at 0.067λ based on the MIT experiment using two small resonant multi-turn coils under coupled mode operation (Kurs et al., 2007). It was also shown that high antenna radiation efficiency is needed to maximize the power transfer beyond the coupled mode region. This implies that existing knowledge on the design of highly efficient small antennas for far-field applications can be leveraged upon to achieve the upper PTE bound in the near-field region. The design of electrically small but highly efficient antennas is a research topic that has been well investigated in the past decade (Dobbins & Rogers, 2001; Choo & Ling, 2003; Best, 2004; Lim & Ling, 2006). Among the various small antenna designs, the most well known is the folded spherical helix (FSH) by Best (Best, 2004). In his design, four arms are wound helically along a spherical surface and the arms are connected at the top. Such design uses multiple folds to step up the radiation resistance. The reported size of kr is 0.38 (where k is the free-space wave number and r is the minimum size of the imaginary sphere enclosing the antenna). The FSH monopole antenna yields a low Q (32) and high radiation efficiency η (98.6%) despite its small size. The folded cylindrical helix (FCH) has also been investigated (Johnston & Haslett, 2005; Best, 2009; Best, 2009). Although it does not achieve as low a Q-factor as that of the FSH, it has a form factor that is easier to construct and handle.

While recent research on wireless power transfer has been mostly centered on the coupled mode phenomenon, we focus our attention in this chapter on the near-field region beyond the coupled mode resonance region. In particular, it will be shown that electrically small antennas can be designed to realize efficient wireless power transfer in the near field. This chapter is organized as follows. In Sec. 2, we show that the theoretical bound derived in (Lee & Nam, 2010) can be approached in practice by the use of two electrically small but highly efficient FCH dipoles. In Sec. 3, we discuss transmitter and receiver diversity as a means to extend the range or efficiency of the near-field power transfer. We derive the theoretical PTE bounds under transmitter diversity and receiver diversity and then achieve the bounds experimentally using actual FCH dipoles. In Sec. 4, we investigate electrically small, directive antennas as a means of increasing the range or power transfer efficiency in the near-field region. Sec. 5 presents conclusions and discusses future research directions.

2. Achieving power transfer bound using electrically small antennas

2.1 Theoretical bound for near-field power transfer and antenna design considerations

We will first review the theoretical results from (Lee & Nam, 2010) on the theoretical PTE bound between two small antennas in the near-field region. The PTE is defined as the ratio of the power dissipated in the load of the receive antenna to the input power accepted by the transmit antenna:

$$PTE = \frac{R_{load}\left|I_2\right|^2 / 2}{R_{in}\left|I_1\right|^2 / 2} \qquad (1)$$

where R_{load} is the load resistance, I_2 is the current at the load of the receiving antenna, R_{in} is the input resistance and I_1 is the current at the feed point of the transmitting antenna. To derive the maximum possible PTE for small antennas, (Lee & Nam 2010) assumed that the field radiated by a small electric dipole antenna can be expressed in terms of the TM_{10} spherical harmonic. Using the addition theorem of spherical wave functions, the radiated field from the transmit antenna can be re-expressed in terms of impinging spherical modes on the receive antenna. The strength of the resulting inward-traveling TM_{10} mode onto the receive antenna can thus be found. From this field-based formulation, a two-port description of the transmit-receive antenna system is obtained. Next, they assumed that the optimal load to achieve maximum power transfer, or the so-called Linville load (Balanis, 1997), is used on the receive antenna. The final maximum PTE bound takes on the following simple, closed-form expression:

$$PTE = \frac{|T|^2}{2 - \mathrm{Re}\left[T^2\right] + \sqrt{4\left(1 - \mathrm{Re}\left[T^2\right]\right) - \mathrm{Im}\left[T^2\right]^2}}$$

$$T = \eta\frac{3}{2}\cdot\left[-\sin^2\theta\frac{1}{jkd} + \left(3\cos^2\theta - 1\right)\cdot\left\{\frac{1}{\left(jkd\right)^2} + \frac{1}{\left(jkd\right)^3}\right\}\right]\cdot e^{-jkd}$$

(2)

In the above expression, η is the radiation efficiency of the receiving antenna, θ is the angle of the receiving antenna with respect to the transmitting antenna (see inset to Fig. 2), $k=2\pi/\lambda$ is the free-space wave number and d is the spacing between the antennas. The same bound also holds true for a small magnetic dipole, or a small loop antenna, as the excitation of the TE_{10} mode leads to the same final expression.

The PTE bounds vs. distance (measured in wavelengths) are computed using (2) and plotted in Fig. 2 for different η and θ values. It is observed that higher PTE is achieved when the two antennas are in the co-linear configuration ($\theta=0$) than when they are in the parallel configuration ($\theta=\lambda/2$) up to a distance of 0.4λ. This is clearly a feature unique to the near-field region. Beyond this distance, our usual far-field intuition takes hold, i.e., antennas couple much more strongly in the parallel instead of the co-linear configuration. Several additional antenna design implications can be inferred from Fig. 2. First, we observe that PTE decreases rapidly as η decreases, implying that antennas with high radiation efficiency are needed to achieve the best results. It is crucial that any dissipative losses in the antenna are small relative to the radiated power. For example, simply using very short dipoles will not be a good choice to realize efficient power transfer since they have small radiation resistance and thus poor radiation efficiency when constructed using real conductors. Another important consideration is impedance matching. The PTE bounds in Fig. 2 are derived by assuming a different optimal load value is used at every distance. Any deviation from this optimal load will critically lower the realizable PTE. In the far field, the optimal load is simply the conjugate of the input impedance of the receiving antenna. In the coupled mode region, this conjugate matching condition is strongly violated. Fortunately, the conjugate matching condition is only weakly perturbed in the near-field region, since the antennas are not strongly coupled. Therefore, enforcing the conjugate matching condition leads to a good design choice for power transfer in the near field, without the need for a complex matching network that is a function of antenna separation. If a 50Ω load is desirable from a practical point of view, the receiving antenna can be designed to have an input impedance of 50Ω to approach the PTE bound. The same antenna design, when used for the transmit antenna, will also present a convenient impedance for the transmitter circuitry.

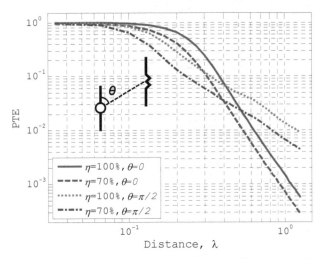

Fig. 2. PTE bound versus distance for different radiation efficiencies and antenna orientations. ©2010 IEEE (Yoon & Ling, 2010).

2.2 Design of electrically small, folded cylindrical helix (FCH) antennas

To approach the theoretical PTE bound using electrically small antennas, we need to design highly efficient antennas with a 50Ω input impedance. As discussed in Sec. 1, the folded cylindrical helix (FCH) is one such candidate with a good form factor and a well-understood design methodology. Since the radiation resistance of an electrically small antenna drops quadratically as a function of its height, the FCH design uses the folding concept to step up the small radiation resistance to improve the radiation efficiency and provide good matching. When N multiple folding arms are used for a highly symmetrical antenna structure, equal currents are induced on all the arms, resulting in a radiation resistance (R_{rad}) that scales approximately as N^2 while the loss resistance (R_{loss}) is increased only by a factor of N, thus leading to a high radiation efficiency (Lim & Ling, 2006).

Fig. 3 shows the designed FCH dipole operating at 200MHz based on 18-gauge copper wires. The diameter and height of the antenna are chosen to be approximately equal and are confined to a maximum dimension of 10.5cm. The antenna fits within a $kr=0.31$ sphere. NEC is utilized in the antenna modeling. The number of turns (1.25-turn) and the number of arms (4 arms) are chosen to reach the input impedance target of 50Ω while achieving a resonant frequency of 200MHz. The resulting antenna has an input resistance of 48.9Ω with a corresponding η of 93% based on NEC simulation.

Two FCH dipoles based on the above design are built and tested. They are constructed by winding copper wires on paper support. During testing, the feed point for each FCH is connected to a 2:1 transformer balun, which is characterized separately. The balun is then de-embedded from the measurement to obtain the S_{11} of each antenna. Fig. 4 compares the simulated and measured input impedances. As can be seen in Fig. 4, the measured results are shifted slightly downward by 5MHz from the simulation. This is due to fact that the wire windings in the built antennas are slightly longer than the design. However, the input resistances are still measured to be 49.1Ω and 48.9Ω, respectively, at their resonant frequencies. This pair of antennas are used as the transmit and receive antennas in the PTE measurement.

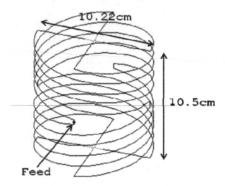

Fig. 3. The designed 1.25-turn, 4-arm folded cylindrical helix dipole. ©2010 IEEE (Yoon & Ling, 2010).

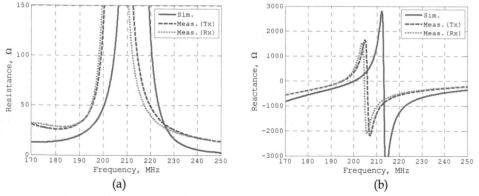

Fig. 4. Simulated and measured input impedance of the folded cylindrical helix dipole. (a) Input resistance. (b) Input reactance. ©2010 IEEE (Yoon & Ling, 2010).

2.3 Near-field power transfer simulation and measurement

The power transfer efficiency between the two designed FCH antennas is simulated using NEC as well as measured. In the simulation, a 50-Ω resistive load is placed on the receive FCH. In this manner, any potential impedance mismatch between the receiver and the load is reflected in the resulting PTE value. The center-to-center distance between the two FCHs is varied from 0.15m to 2m. At very close-in range the resonant frequency of each antenna splits into two, which is consistent with the even and odd mode behavior in the coupled mode region (Kim & Ling, 2007). For the measurement, the FCH dipoles are mounted on 2m-high tripods and measured using a vector network analyzer. Fig. 5 shows the photos for the outdoor measurement setup. PTE is calculated from the measured S-parameters as $|S_{21}|^2/(1-|S_{11}|^2)$. The two baluns are again de-embedded to obtain the S_{21} between the two antennas. Both the co-linear and parallel configurations are simulated and measured.

(a)   (b)

Fig. 5. Photos of the outdoor measurement setup. Each FCH is connected to a 2:1 transformer balun. (a) $\theta=0$. (b) $\theta=\pi/2$. ©2010 IEEE (Yoon & Ling, 2010).

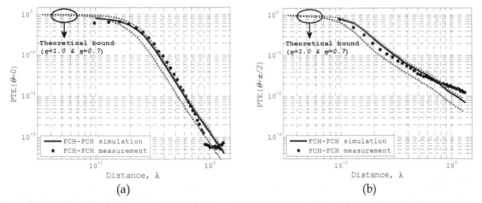

(a)   (b)

Fig. 6. Simulated and measured PTE of two FCH-FCH dipoles. (a) $\theta=0$. (b) $\theta=\pi/2$. ©2010 IEEE (Yoon & Ling, 2010).

Fig. 6 shows the simulated (solid lines) and measured (dots) PTE results versus distance in wavelength. The PTE bounds from Eq. (2) shown earlier in Fig. 2 are re-plotted as dotted-lines for comparison. We observe that the simulated FCH-FCH results nearly approach the theoretical PTE bound with $\eta=100\%$. The minor difference between them can be attributed to the slightly less than ideal radiation efficiency of the real FCH antenna ($\eta=93\%$) and the small mismatch between the receive antenna and the load. We also observe from Fig. 6a that the simulated PTE curve begins to deviate more from the $\eta=100\%$ reference when the two antennas are very close to each other. This deviation is due to the increasing difference between the optimal Linville load Z_{Linv} and the resistive 50Ω load when the antennas begin to couple tightly at closer distances. The measured FCH-FCH results (shown as dots in both Figs. 6a and 6b) follow the simulation results fairly well. The discrepancy is likely caused by the non-negligible physical size of the balun boxes and the slight misalignment during the measurement. Overall, the measurement results clearly demonstrate that practical electrically small antennas can be designed and realized to approach the theoretical bounds derived in (Lee & Nam, 2010) for wireless power transfer beyond the coupled mode region. The measured results showed a PTE of 40% at a distance of 0.25λ in the co-linear configuration, very close to the theoretical bound. For comparison, the meander monopole antennas reported in (Kim & Ling, 2007) showed a 40% PTE at a much shorter distance of 0.044λ (0.31m at 43MHz) in the parallel configuration. This section highlights the role of the theoretical bound in guiding antenna design. It also underscores the importance of antenna design in achieving highly efficient power transfer in the near field. In the next two sections, we explore possible ways to further extend the range or efficiency of the power transfer.

3. Power transfer enhancement using transmitter diversity and receiver diversity

One way to extend the range or efficiency of the power transfer in the near-field region is to use receiver and/or transmitter diversity (see Fig. 7). Multiple receiver scenarios have been studied in the coupled mode region to power multiple devices simultaneously from a single source (Cannon et al., 2009; Kurs et al., 2010). Multiple transmitting and receiving coils have also been investigated for device charging at very close range (Casanova et al, 2009). Here, we focus our attention on transmitter and receiver diversity in the near-field region.

Fig. 7. Diversity configuration. (a) Transmitter diversity. (b) Receiver diversity.

3.1 Near-field power transfer bound under transmitter diversity
To begin, we define the PTE for multiple transmitters and a single receiver as the ratio of the power dissipated in the load to the total power accepted by the transmit antennas. For the two transmitter case, this is given as:

$$PTE = \frac{\text{Re}\{Z_{load}\} \cdot |I_3|^2 / 2}{\text{Re}\{V_1 \cdot I_1^*\} / 2 + \text{Re}\{V_2 \cdot I_2^*\} / 2} \tag{3}$$

where Z_{load} is the load impedance, I_3 is the current at the load of the receiving antenna (port 3), V_1, V_2 are the input voltages and I_1, I_2 are the currents at the feed points of the transmitting antennas (ports 1 and 2). With specified input voltages at the transmitter ports and a given load, the terminal currents can be obtained once the 3-by-3 Z-matrix of the network is known:

$$\begin{bmatrix} Z_{11} & Z_{12} & Z_{13} \\ Z_{21} & Z_{22} & Z_{23} \\ Z_{31} & Z_{32} & Z_{33} \end{bmatrix} \begin{bmatrix} I_1 \\ I_2 \\ I_3 \end{bmatrix} = \begin{bmatrix} V_1 \\ V_2 \\ -I_3 \cdot Z_{load} \end{bmatrix} \tag{4}$$

To obtain the Z-matrix, we can extend the mutual impedance expression between two electrically small antennas derived by Lee and Nam (Lee & Nam, 2010) to the multiport case as follows:

$$Z_{mn} = \begin{cases} Z_a, & m = n \\ \mathrm{Re}[Z_a] \cdot T, & m \neq n \end{cases} \tag{5}$$

In Eq. (5), we approximate the self impedance by Z_a, the stand-alone impedance of a small dipole, which can be calculated using the induced EMF method (Balanis, 1997). T is described in Eq. (2). To compute the upper PTE bound, we need to find the optimal value for Z_{load}, Z_{opt}. Unfortunately, the closed form solution (i.e., the Linville load) used in Sec. 2 is only available for a two-port network. For three ports or more, a numerical optimization must be performed. Based upon (3) through (5), a local search for Z_{opt} to reach the maximum PTE is carried out with the antenna configuration shown in Fig. 7(a). The separation of the transmitters is fixed as D and a single receiver is moved between them. The antennas are in the parallel configuration ($\theta = \pi/2$) and the radiation efficiencies are set to 100%. To compute the bound, the optimal load is found numerically at every receiver position, with the conjugate value of the input impedance of the receiving antenna as the initial guess. The same input voltages are assumed ($V_1 = V_2$) at the two transmitters. For Z_a, we use a value of 0.079-11270j [Ω], which is the input impedance for a $\lambda/50$ dipole computed by the induced EMF method.

Figs. 8a and 8b show the maximum PTE bounds derived in this manner when the transmitters are 0.7λ and 1.0λ apart, respectively. The theoretical PTE bound for the single transmitter case is also plotted by dashed lines for reference in both figures. In Fig. 8a, it is interesting to see that a very stable PTE region can be created as a function of the receiver position d. A PTE of 22% is maintained over a 0.2λ–0.5λ region between the two transmitters. This is also significantly improved from that of the single transmitter case. In this region, the fields due to the two transmitters add constructively since the receiver distance from the two transmitters is small. However, at larger spacing between the transmitters, rippling starts to appear due to the expected standing wave interference. This is shown in Fig. 8b for $D = 1.0\lambda$.

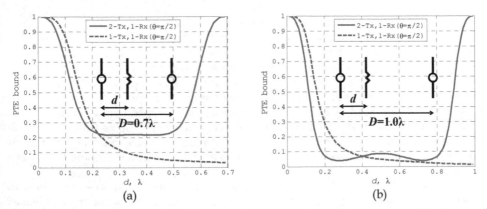

Fig. 8. PTE bound for transmitter diversity (two transmitters and one receiver) in the parallel configuration. The two transmitters are separated by the fixed distance D. (a) $D=0.7\lambda$. (b) $D=1.0\lambda$. ©2011 IEEE (Yoon & Ling, 2011).

To realize this PTE bound under transmitter diversity in practice, impedance matching and antenna radiation efficiency need to be addressed. Z_{opt} for the PTE bound calculation in Fig. 8a ($D=0.7\lambda$) is plotted in Fig. 9. For reference, $Z_a=0.079-11270j$ [Ω]. It is observed that Z_{opt} is also rather stable in the flat PTE region of 0.2λ–0.5λ. In particular, its numerical value is close to the conjugate of the input impedance of the stand-alone antenna. An important implication of this observation is that if we want to use a constant 50-Ω load, then it is acceptable to design the receiving antenna to have an input impedance of 50Ω to approach the upper PTE bound in the region. In terms of radiation efficiency of the antennas, it is observed from simulation that the PTE decreases significantly as the radiation efficiency is reduced. This is the same as the conclusion reached earlier in Sec. 2. Therefore, the antennas should be designed to have radiation efficiency as close to 100% as possible.

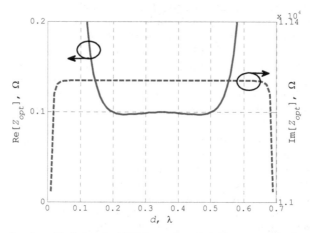

Fig. 9. Optimal load value (Z_{opt}) for the PTE bound with 0.7λ separation between the transmitters (Fig. 8a case), with respect to $Z_a=0.079-11270j$ [Ω]. ©2011 IEEE (Yoon & Ling, 2011).

3.2 Power transfer simulation and measurement

We set out to demonstrate through simulation and measurement that the derived PTE bound can be approached using actual antennas. The same electrically small FCH dipoles designed in Sec. 2 are used. The antenna has an input resistance of 48.9Ω with a corresponding radiation efficiency of 93% and fits inside a $kr=0.31$ sphere at 200MHz. Three identical FCH dipoles are used in the NEC simulation and the setup is shown in Fig. 10. Two FCH dipoles are used as transmitting antennas with a separation of 1.05m (0.7λ at 200MHz, as measured between the two antenna ports), and an FCH dipole with a fixed 50-Ω resistive load is used as the receiving antenna. The position of the receiving antenna is changed between the transmitters from 0.15m to 0.90m. PTE is calculated using Eq. (3) at a fixed frequency of 200MHz.

For measurement, three FCH dipoles are first fabricated and measured. They are tuned to have the same resonance frequency. The measured input resistances of the three antennas are 61.7 Ω, 63.5 Ω and 55.3 Ω at 194.5 MHz. During the PTE measurement, the three FCH dipoles are mounted on 2m high tripods to minimize ground effects and measured using a vector network analyzer. The two transmitting FCH dipoles are separated by 1.08m (0.7λ at

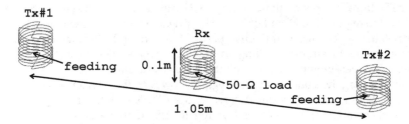

Fig. 10. NEC simulation setup for the two-transmitter and one-receiver scheme using electrically small folded cylindrical helix dipoles. The antennas are in the parallel configuration. ©2011 IEEE (Yoon & Ling, 2011).

194.5MHz) and the receiving one is moved from 0.15m to 0.90m. The 3-by-3 scattering matrix of the antennas at each position is obtained by three sets of measurements. In each set, one antenna is terminated by a 50-Ω load while the other two are connected through the baluns to the ports of the vector network analyzer. The baluns are characterized and de-embedded to obtain the S-parameters. The PTE is calculated from the measured S-parameters as:

$$PTE_{meas} = \frac{|S_{31} + S_{32}|^2}{1 - |S_{11} + S_{12}|^2 + 1 - |S_{22} + S_{21}|^2} \tag{6}$$

where ports 1 and 2 are for the transmit antennas and port 3 is for the receive antenna.
The simulated and measured PTE ($D=0.7\lambda$) versus the position of the receiving antenna is shown in Fig. 11. The derived theoretical PTE bound is re-plotted as a solid line for reference. The simulated PTE values using the FCH dipoles are plotted as red circles. They approach the theoretical PTE bound between 0.2λ and 0.5λ, showing a stable 20% PTE region. The slightly lower PTE from the theoretical bound is due to the imperfect radiation efficiency of the designed FCH dipoles and a small load mismatch. It is interesting to see

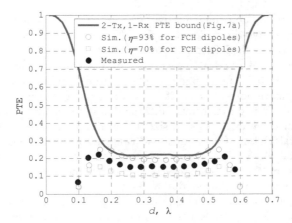

Fig. 11. Simulated and measured PTE for transmitter diversity using three FCH dipoles. ©2011 IEEE (Yoon & Ling, 2011).

that the PTE drops sharply as the receiving FCH dipole moves close to each transmitting antenna (i.e. the regions $d<0.2\lambda$ and $d>0.5\lambda$). This phenomenon can be explained by the large load mismatch. Fig. 9 shows that the real part of the optimal load value deviates from the center region due to strong coupling as the receiving antenna comes close to either transmitter. The measurement result is plotted as black dots and it follows the trend of the simulation, showing the same stable PTE region. However, the PTE level is about 3% lower than the simulation. To explain this difference, the PTE is simulated using the same FCH dipoles but their radiation efficiency is lowered to 70%. It shows an even lower PTE level than the measurement. A Wheeler cap measurement is also performed for the fabricated FCH dipoles and the measured radiation efficiencies are between 80% and 90%. Therefore, we conclude that the lower radiation efficiencies of the built antennas are the main cause of the lower PTE values as compared to the simulation. This again highlights the importance of high antenna efficiency for achieving efficient near-field power transfer.

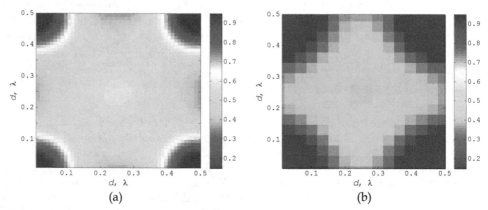

Fig. 12. Four-transmitter and one-receiver scheme. (a) PTE bound. (b) NEC simulation using electrically small folded cylindrical helix dipoles. ©2011 IEEE (Yoon & Ling, 2011).

Finally, the two-transmitter and single-receiver case is extended to four transmitters and a single receiver over a two-dimensional region. The same method used for the two transmitter case is used for the PTE derivation except that a 5-port network is considered. The four transmitters are located at the corners of a square region. The length of the diagonal region is set to 0.7λ and a single receiver is moved within the region. Again the optimal load for maximum power transfer is found through a numerical search. The derived PTE bound is plotted as a two-dimensional intensity plot in Fig. 12a. It is seen that a stable PTE of 50% is generated within a region of size 0.3λ at the center. Next, actual FCH dipoles are used in the PTE calculation. The optimal load value at each position is also replaced by a fixed 50-Ω load at the receive FCH dipole and the result is shown in Fig. 12b. The same trends as the two transmitter case are observed. First, when the receiver comes closer to each transmitter, the PTE again falls sharply due to load mismatch. Second, despite the slightly lower value (41%) from the PTE bound due to the imperfect radiation efficiency and load mismatch, the stable PTE region is created along the center region. Thus transmitter diversity can potentially be applied to provide a stable service region for one (or more) mobile receiver(s).

3.3 Near-field power transfer under receiver diversity

The PTE bound for receiver diversity can also be derived by using the same approach for the transmitter diversity. For two receivers, only the PTE definition is different and is given as:

$$PTE = \frac{Re\{Z_{load,2}\} \cdot |I_2|^2 / 2 + Re\{Z_{load,3}\} \cdot |I_3|^2 / 2}{Re\{V_1 \cdot I_1^*\} / 2} \tag{7}$$

where port 1 is the transmitter and ports 2 and 3 are the receivers. $Z_{load,2}$ and $Z_{load,3}$ are the load impedances.

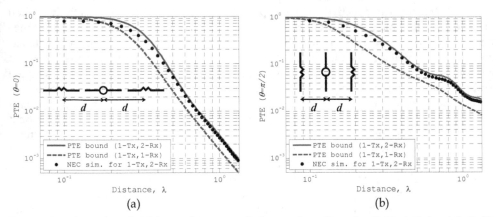

Fig. 13. PTE bound and NEC simulation result for receiver diversity (1-Tx, 2-Rx). (a) $\theta=0$. (b) $\theta = \pi/2$.

Fig. 13 shows the PTE bounds for the co-linear and parallel configurations. In these plots, the distances between the transmitter and the two receivers are set to be equal to simplify the calculation. The result of this constraint is that the optimal load impedances at the two receiving antennas are the same, thus simplifying the numerical search. The efficiencies of the antennas are set as 100%. The PTE bounds for the single transmitter and single receiver case are plotted in dashed lines for reference. As expected, the PTE bound is extended under receiver diversity. However, some rippling of the PTE curve can be seen due to coupling of the antennas. NEC simulation using three FCH dipoles (R_{in}=48.9Ω, η=93% and kr=0.31) with fixed 50-Ω loads is also carried out and the result is plotted as dots in Fig. 13. Not surprisingly, it follows the PTE bound well as it did in the single-transmitter single-receiver scenario.

In this section, we have derived the upper bounds for power transfer under the transmitter diversity and receiver diversity scenarios. It was also shown that such bounds can be approached by using highly efficient electrically small antennas. Generalization to the case of multiple transmitters and multiple receivers is reported in (Jun, 2011).

4. Power transfer enhancement using small directive antennas

Another possible way to enhance the range or efficiency of near-field wireless power transfer is to use spatial focusing antennas. While the role of directivity is very clearly described by the Friis transmission formula in the far field, spatial focusing in the near field is not as well understood (Schantz, 2005; Kim et al., 2010). In the derivation of the upper bound for near-field power transfer in (Lee & Nam, 2010), only the lowest TM_{10} or TE_{10} mode was considered. In this section, we investigate whether small directive antennas can be used as a means of increasing the range or PTE of near-field power transfer.

4.1 Feasibility and antenna design considerations

We first test the feasibility of near-field coupling enhancement using directive antennas designed for far-field application. Uzkov showed that the end-fire directivity of a periodic linear array of N isotropic radiators can approach N^2 as the spacing between elements decreases, provided the magnitude and phase of the input excitations are properly chosen (Uzkov, 1946). Such a directivity value represents the so-called "superdirectivity" when compared to the maximum attainable directivity for isotropic elements spaced half-wavelength apart, especially because the directivity increases as the length of the linear array becomes smaller (Altshuler et al., 2005). Thus, the directivity of a two-element array of isotropic radiators would approach a value of four, i.e., 6 dB higher than that of a single isotropic radiator. Parasitic implementation of superdirectivity is also feasible. For example, a two-element, 0.02λ spacing Yagi antenna with a driver and a reflector each about half-wavelength (driver: 0.4781λ and reflector: 0.49λ) shows a directivity of 7.2dB in the far field (Lim & Ling, 2006). This configuration is chosen for both the transmit and receive antennas in our near-field PTE simulation. Under no conductor losses and with the optimal load (Z_{Linv}) for maximum coupling at every distance, the PTE between two such antennas is calculated. We shall call this particular antenna configuration under no conductor loss and with the optimal load an "idealized superdirective array" in subsequent discussions.

The calculated PTE is shown by a solid line in Fig. 14. PTE enhancement is observed from 0.1λ as compared to the PTE bound derived for small antennas, which is plotted by a dashed line. At far distances, the enhancement approaches about 11 dB, which is well predicted by the Friis far-field formula (2*(7.2dB-1.76dB)). While the solid line in Fig. 14 shows an impressive improvement in PTE performance when directive antennas are used, there are practical implementation issues. First, when the optimal load is replaced by a fixed 50-Ω

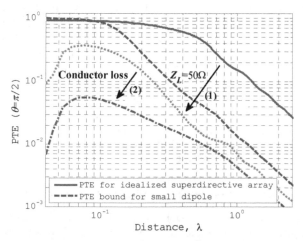

Fig. 14. PTE enhancement in the near-field region using an idealized superdirective array.

load, the PTE degrades sharply, as marked by the first arrow in Fig. 14. Second, the PTE degrades even more if conductor loss is considered, as marked by the second arrow in Fig. 14. These degradations are similar to the well-known pitfalls in realizing superdirective antennas for far-field applications, namely, an idealized superdirective array often comes with a large impedance mismatch and low radiation efficiency when implemented in practice. Therefore, to capture the benefit of high directivity for near-field power transfer, one must pay careful attention to antenna design. First, the radiation efficiency of the antenna should be high to minimize the conductor loss effect. Second, the input impedance of the antenna needs to be close to 50Ω for impedance matching to a standard 50-Ω resistive load. Finally, it is desirable that the antenna size be as small as possible.

4.2 Design of an electrically small FCH Yagi

We present in this section an electrically small Yagi antenna designed based on the FCH dipole to achieve good directivity, radiation efficiency and impedance matching. The antenna is comprised of two elements, a driver and a reflector. Each element is an FCH structure that has been thoroughly discussed in previous sections. The dimensions of the antenna including the height and the radius of each element, the spacing between the two elements and the number of arms are optimized using a local optimizer in conjunction with NEC. The far-field realized gain, which accounts for directivity, radiation efficiency and matching, is chosen to be the objective function.

The optimized antenna design and a photo of the fabricated antenna are shown in Fig. 15. The antenna feed is located at the center of one arm in the driver. The wire length of one arm in the driver is 85 cm (0.57λ at 200MHz) and that of the reflector is slightly longer at 92 cm (0.61λ at 200MHz). The entire antenna fits inside a kr=0.95 sphere, where r is the radius of the imaginary sphere that encloses the entire antenna. A copper wire radius of 0.5mm (18-AWG) is chosen for both elements. The optimized Yagi antenna has an input resistance of 50.9Ω at resonance with a corresponding radiation efficiency of 96%. The maximum simulated directivity, gain and realized gain in the forward direction at 200MHz are 6.76, 6.61 and 6.60dB, respectively. The front-to-back ratio is 12.3dB.

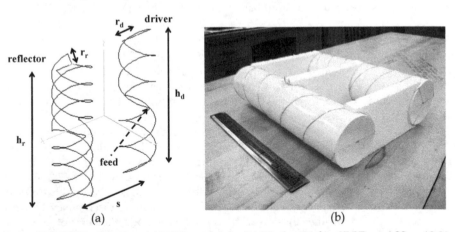

Fig. 15. Small FCH Yagi design. (a) NEC model: h_r=34.74; r_r=4.75; h_d=35.37; r_d=4.28; s=19.90 (units in centimeters); number of arms=three for reflector, two for driver; number of turns=1.25 for each arm. (b) Photo of the fabricated antenna. ©2011 Wiley (Yoon & Ling, 2011).

A prototype of the antenna in Fig. 15(a) is built by winding copper wires on paper support. The position of the reflector and driver parts is held in place by Styrofoam. The two small metal tips in the driver are connected to a 2:1 transformer balun in the measurement. The balun is characterized and de-embedded to obtain the S_{11} of the antenna. Fig. 16(a) shows the simulated and measured input impedances. Other than a slight downward shift of 5MHz, the measured results show good agreement with the simulation. The input resistance is measured as 56.0Ω at its resonant frequency of 195.3MHz. The simulated and measured realized gain and front-to-back ratio are plotted in Fig. 16(b). The measured realized gain is 6.2dB at 195.3MHz, which is lower than the simulated realized gain by 0.4dB. Minor RF interference noises can be observed at 175, 215, and 225MHz due to the outdoor environment. The front-to-back ratio is measured as 14.5dB at 195.3MHz, which is slightly higher than the simulated value of 12.3dB. The peaks above the resonant frequencies in both the measurement and simulation are due to a null in the backward direction.

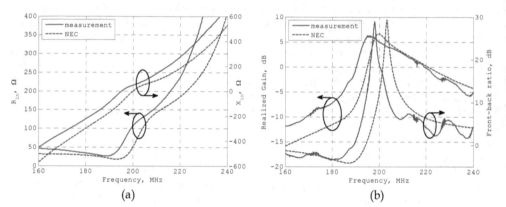

(a)   (b)

Fig. 16. Simulation and measurement results from the designed FCH Yagi antenna. (a) Input impedance. (b) Realized gain and front-back ratio. ©2011 Wiley (Yoon & Ling, 2011).

4.3 Power transfer simulation and measurement

PTE values are simulated and measured using the two designed FCH Yagi antennas. A 50-Ω resistive load is placed on the receiving antenna in the NEC simulation. The distance between the antennas is measured as that between the centers of the two antennas. The photos for the outdoor measurement setup are shown in Fig. 17. The PTE is calculated from the measured S-parameters as $|S_{21}|^2/(1-|S_{11}|^2)$.

The simulated PTE results with the two FCH Yagi antennas with a fixed 50-Ω resistive load are shown as solid dots in Fig. 18. The PTE calculated with the idealized superdirective array (i.e. without conductor loss and with optimal load at every distance) in Fig. 14 is re-plotted for reference. The dashed line is the Friis far-field formula with two 7.2dB antennas. It is observed that the simulated PTE between the FCH Yagis follows the trend of the idealized superdirective array for spacing greater than 0.3λ. The discrepancy between the two is mainly caused by the lower directivity and radiation efficiency of the designed antennas as compared to the idealized superdirective array. When the two Yagi antennas are spaced less than 0.3λ, the PTE between them is rather degraded. This is due to the strong coupling between the antennas, which causes the current distribution on each Yagi antenna to deviate from its stand-alone design for high directivity. The measured results are plotted as open circles and they show reasonable agreement with the simulation. Both the simulation and measurement show that 40% PTE is achieved at a distance of 0.45λ based on the center-to-center distance definition.

(a)   (b)

Fig. 17. Measurement setup. (a) FCH Yagi-FCH Yagi. (b) FCH Yagi-FCH dipole.

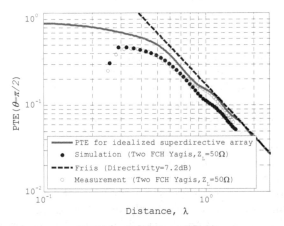

Fig. 18. Simulated and measured PTE for FCH Yagi-FCH Yagi.

To see the improvement in PTE by using directive antennas in the near-field region, the PTE results in Fig. 18 are compared to those from using FCH Yagi (Tx)-FCH dipole (Rx) and FCH dipole (Tx)-FCH dipole (Rx) in Fig. 19. The simulated values are plotted as lines and the corresponding measurement results are plotted as dots. At 0.45λ, the PTE difference between the FCH Yagi-FCH Yagi coupling and FCH dipole-FCH dipole coupling is 9.11 dB, showing good enhancement, though it is slightly lower than the 9.7 dB in the far field. This statement is also supported by the measurement results, which follow the simulation results fairly well. From this study, we see that the far-field realized gain can be used as a good surrogate for designing small directive antennas for near-field power transfer. This is an important practical consideration, since using PTE directly as the cost function is

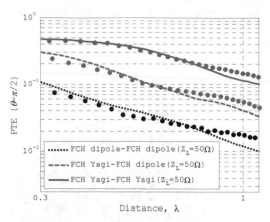

Fig. 19. PTE comparison. FCH dipole-FCH dipole, FCH Yagi-FCH dipole and FCH Yagi-FCH Yagi.

computationally more expensive and can lead to different designs at different distances. However, we also note that the PTE improvement comes at the price of antenna size, as the designed FCH Yagi has about a three-fold increase in kr as compared to that of the FCH dipole.

5. Conclusion

The use of the coupled mode resonance phenomenon for non-radiative wireless power is being studied intensively and many works have been reported. However, the distance over which such phenomenon exists is very short when measured in terms of wavelength. In this chapter, we have focused our attention on the near-field region beyond the coupled mode resonance region as a means of efficient wireless power transfer. First, the theoretical power transfer efficiency (PTE) bound was demonstrated by the design of electrically small, highly efficient folded cylindrical helix (FCH) dipole antennas. A 40% PTE was achieved at the distance of 0.25λ between the antennas in the co-linear configuration. To further extend the range or efficiency of the power transfer, transmitter diversity and receiver diversity were investigated. For transmitter diversity, it was found that a stable PTE region can be created when multiple transmitters are sufficiently closely spaced. Subsequently, such a stable PTE region was demonstrated using electrically small FCH dipoles. The measurement results highlighted the importance of maintaining high antenna radiation efficiency in realizing efficient wireless power transfer. For receiver diversity, it was found that the PTE can also be improved as the number of the receivers is increased. Finally, we investigated whether small directive antennas can be used as a means of enhancing near-field wireless power transfer. It was found that the range of efficiency can indeed be enhanced by using small directive antennas. It was also shown that the far-field realized gain is a good surrogate for designing small directive antennas for near-field power transfer.

To conclude this chapter, we shall mention several potential research topics that we believe are important for the practical implementation of near-field power transfer. First, the effects of surrounding environments on the near-field coupling should be carefully examined. One of the attractiveness of magnetic resonant coupling is that dielectric materials have only a marginal effect on such mode of power transfer. We expect stronger material effects in the near field due to the presence of both electric and magnetic fields around self-resonant antennas. The interactions of the near fields with materials need to be quantified. Second, it would be interesting to study whether an optimal frequency exists for maximizing the range or efficiency of the transfer. 200MHz was chosen for all of the examples in this chapter as a matter of convenience in our measurements. A lower frequency would naturally lead to a longer absolute range in the power transfer. However, maintaining high efficiency in electrically even smaller antennas also becomes more challenging. Therefore, an optimal frequency may exist by considering the different practical implementation constraints (e.g., wire radius, antenna size). Some work has recently been reported along this line in (Poon et al., 2010). Finally, the design of orientation independent antennas is another topic that would be of interest in servicing mobile users.

6. References

Altshuler, E. E.; O'Donnell, T. H., Yaghjian, A. D. & Best, S. R. (Aug. 2005). A monopole superdirective array, *IEEE Trans. Antennas Propag.*, vol. 53, no. 8, pp. 2653–2661

Balanis, C. A. (1997). *Antenna Theory: Analysis and Design*, 2nd ed., John Wiley & Sons

Best, S. (Apr. 2004). The radiation properties of electrically small folded spherical helix antennas, *IEEE Trans. Antennas Propag.*, vol. 52, no. 4, pp. 953–960

Best, S. (Mar. 2009). A comparison of the cylindrical folded helix Q to the Gustafsson limit, in *Proc. EuCAP*, Berlin, Germany, pp. 2554–2557

Best, S. (Dec. 2009). The quality factor of the folded cylindrical helix, *Radioengineering*, vol. 18, pp. 343–347

Brown, W. C. (Sep. 1984). The history of power transmission by radio waves, *IEEE Trans. Microw. Theory Tech.*, vol. 32, pp. 1230–1242

Brown, W. C. (Jun. 1992). Beamed microwave power transmission and its application to space, *IEEE Trans. Microw. Theory Tech.*, vol. 40, pp. 1239–1250

Bolomey, J. C.; Capdevila, S., Jofre, L. & Romue, J. (2010). Electromagnetic modeling of RFID-modulated scattering mechanism. Application to tag performance evaluation, *Proc. IEEE*, vol. 98, no. 9, pp. 1555-1569

Cannon, B. L.; Hoburg, J. F., Stancil, D. D. & Goldstein, S. C. (Jul. 2009). Magnetic resonant coupling as a potential means for wireless power transfer to multiple small receivers, *IEEE Trans. Power Electronics*, vol. 27, pp. 1819–1825

Casanova, J. J.; Low, Z. N. & Lin, J. (Aug. 2009). A loosely coupled planar wireless power system for multiple receivers, *IEEE Trans. Industrial Electronics*, vol. 56, pp. 3060–3068

Choo, H and Ling, H. (Oct. 2003). Design of electrically small planar antennas using an inductively coupled feed, *Elect. Lett.*, vol. 39, pp. 1563-1564

Conner, M. (July 5, 2007). Wireless power transmission: no strings attached, *Electronic Design News*

Dobbins, J. A. & Rogers, R. L. (Dec. 2001). Folded conical helix antenna, *IEEE Trans. Antennas Propag.*, vol. 49, no. 12, pp. 1777–1781

Eom, K. & Arai, H. (Mar. 2009). Wireless power transfer using sheet-like waveguide, in *Proc. EuCAP*, Berlin, Germany, pp. 3038–3041

Finkenzeller, K. (2003). *RFID Handbook: Fundamentals and Applications in Contactless Smart Cards and Identification*, 2nd ed. Chichester, England; Hoboken, N.J.: Wiley

Fotopoulou, K. & Flynn, B. W. (Oct. 2006). Wireless powering of implanted sensors using RF inductive coupling, in *Proc. IEEE Conf. Sensors*, Daegu, Korea, pp. 765–768

Fotopoulou, K. & Flynn, B. W. (Mar. 2007). Optimum antenna coil structure for inductive powering of passive RFID tags, in *Proc. IEEE Intl. Conf. on RFID*, pp. 71–77, Grapevine, TX

Fotopoulou, K. & Flynn, B. W. (Feb. 2011). Wireless power transfer in loosely coupled links: coil misalignment model, *IEEE Trans. Magn.*, vol. 47, no. 2, pp. 416–430

Glaser, P. E. (1968). Power from the sun: its future, *Science*, vol. 162, pp. 857–886

Goubau, G. (1970). Microwave power transmission from an orbiting solar power station, *J. Microw. Power*, vol. 5, pp. 223–231

Hansen, J. E. (1988). Spherical near-field antenna measurements, *IEE Electromagnetic Waves Series 26*

Imura, T.; Okabe, H. & Hori, Y. (Sep. 2009). Basic experimental study on helical antennas of wireless power transfer for Electric Vehicles by using magnetic resonant couplings, in *Proc. IEEE VPPC*, Dearborn, MI, pp. 936–940

Jing, H. C. & Wang, Y. E. (Jul. 2008). Capacity performance of an inductively coupled near field communication system, in *IEEE Antennas Propag. Int. Symp. Dig.*, San Diego, CA

Johnston, R. H. & Haslett, J. W. (Jul. 2005). Antennas for RF mote communications, in *IEEE Antennas Propag. Int. Symp. Dig.*, Washington, DC, vol. 4A, pp. 267–270

Jun, B. W. (May 2011). An investigation on transmitter and receiver diversity for wireless power transfer, *MS thesis*, Univ. of Texas at Austin

Jung, Y.-K. & Lee, B. (Jul. 2010). Metamaterial-inspired loop antennas for wireless power transmission, in *IEEE Antennas Propag. Int. Symp. Dig.*, Toronto, ON, Canada

Kim, I.; Xu, S., Schmidt, C. & Rahmat-Samii, Y. (Jul. 2010). On the application of the Friis formula: simulations and measurements, in *Proc. URSI Nat. Radio Sci. Meeting*, Toronto, ON

Kim, Y. & Ling, H. (Nov. 2007). Investigation of coupled mode behavior of electrically small meander antennas, *IET Electronics Letters*, vol. 43, pp. 1250–1252Kim, Y. & Ling, H. (Jul. 2008). On the coupled mode behavior of electrically small antennas, in *Proc. URSI Nat. Radio Sci. Meeting*, San Diego, CA

Kurs, A.; Karalis, A., Moffatt, R., Joannopoulous, J. D., Fisher, P. & Soljacic, M. (Jun. 2007). Wireless power transfer via strongly coupled magnetic resonances, *Sciencexpress*, vol. 317, pp. 83–86

Kurs, A.; Moffatt, R. & Soljacic, M. (Jan. 2010). Simultaneous mid-range power transfer to multiple devices, *App. Phys. Lett.*, vol. 96, pp. 044102-1–044102-3

Lee, J. & Nam, S. (Nov. 2010). Fundamental aspects of near-field coupling small antennas for wireless power transfer, *IEEE Trans. Antennas Propag.*, vol. 58, no. 11, pp.3442–3449

Lim, S. & Ling, H. (Jul. 2006). Design of a thin, efficient, electrically small antenna using multiple folding, *Electron. Lett.*, vol. 42, pp. 895–896

Lim, S. & Ling, H. (2006). Design of a closely spaced, folded Yagi antenna, *IEEE Antennas Wireless Propag. Lett.*, vol. 5, pp. 302–305

McSpadden, J. O.; Little, F. E., Duke, M. B. & Ignatiev, A. (Aug. 1996). An inspace wireless energy transmission experiment, *IECEC Energy Conversion Engineering Conference Proceedings*, vol. 1, pp. 468–473

Mickle, M. H.; Mi, M., Mats, L., Capelli, C. & Swift, H. (Feb. 2006). Powering autonomous cubic-millimeter devices, *IEEE Antennas Propagat. Mag.*, vol. 48, pp. 11-21

Nalos, E. J. (Mar. 1978). New developments in electromagnetic energy beaming, *Proc. IEEE*, vol. 55, pp. 276–289

Pan, S.; Jackson, D. R., Chen, J. & Tubel, P. (May 2009). Investigation of wireless power transfer for well-pipe applications, in *Proc. URSI Nat. Radio Sci. Meeting*, Charleston, SC

Poon, A. S. Y.; O'Driscoll, S. & Meng, T. H. (May 2010). Optimal frequency for wireless power transmission into dispersive tissue, *IEEE Trans. Antennas Propag.*, vol. 58, no. 5, pp. 1739–1750

RamRakhyani, A. K.; Mirabbasi, S. & Chiao, M. (Feb. 2011). Design and optimization of resonance-based efficient wireless power delivery systems for biomedical implants, *IEEE Trans. Biomed. Circuits Syst.*, vol. 5, no. 1, pp. 48–63

Sample, A. & Smith, J. R. (Jan. 2009). Experimental results with two wireless power transfer system, *in Proc. IEEE Radio and Wireless Symp.*, pp. 16-18, San Diego, CA

Schantz, H. G. (Jul. 2005). A near field propagation law & a novel fundamental limit to antenna gain versus size, in *IEEE Antennas Propag. Int. Symp. Dig.*, Washington, DC

Shinohara, N. & Matsumoto, H. (Mar. 1998). Experimental study of large rectenna array for microwave energy transmission, *IEEE Trans. Microwave Theory Techniques*, vol. 46, no. 3, pp. 261–267

Smith, S.; Tang, T. B., Terry, J. G., Stevenson, J. T. M., Flynn, B. W., Reekie, H. M., Murray, A. F., Gundlach, A. M., Renshaw, D., Dhillon, B., Ohtori, A., Inoue, Y. & Walton, A. J. (Oct. 2007). Development of a miniaturised drug delivery system with wireless power transfer and communication, *IET Nanobiotechnology*, vol. 1, no. 5, pp 80–86

Strassner, B. & Chang, K. (2002). A circularly polarized rectifying antenna array for wireless microwave power transmission with over 78% efficiency, *IEEE MTT-S International Microwave Symposium Digest*, pp.1535–1538

Tesla, N. (1914). Apparatus for transmitting electrical energy, U.S. Patent 1119732

Thomas, E. M.; Heebl, J. D. & Grbic, A. (Jul. 2010). Shielded loops for wireless non-radiative power transfer, in *IEEE Antennas Propag. Int. Symp. Dig.*, Toronto, ON, Canada

Uzkov, A. I. (1946). An approach to the problem of optimum directive antennae design," *Comptes Rendus (Doklady) de l'Academie des Sciences de l'URSS*, vol. 53, pp. 35–38

Yaghjian, A. D. (Jan. 1982). Efficient computation of antenna coupling and fields within the near-field region, *IEEE Trans. Antennas Propag.*, vol. 30, no. 1, pp. 113–128

Yates, D. C.; Holmes, A. S. & Burdett, A. J. (Jul. 2004). Optimal transmission frequency for ultralow-power short-range radio links, *IEEE Trans. Circuits Syst. I, Reg. Papers*, vol. 51, no. 7, pp. 1405–1413

Yoon, I.-J. & Ling, H. (2010). Realizing efficient wireless power transfer using small folded cylindrical helix dipoles, *IEEE Antennas Wireless Propag. Lett.*, vol. 9, pp. 846–849

Yoon, I.-J. & Ling, H. (2011). Investigation of near-field wireless power transfer under multiple transmitters, *IEEE Antennas Wireless Propag. Lett.*, vol. 10, pp. 662–665

Yoon, I.-J. & Ling, H. (Jun. 2011). An electrically small Yagi antenna with enhanced bandwidth characteristics using folded cylindrical helix dipoles, *Microwave Optical Tech. Lett.*, vol. 53, pp. 1231–1233

Enhanced Coupling Structures for Wireless Power Transfer Using the Circuit Approach and the Effective Medium Constants (Metamaterials)

Sungtek Kahng
The University of Incheon,
South Korea

1. Introduction

Looking around you in your study or office, you will see electronic goods such as a desktop computer and a printer have wires and power cables tangled and plugged. The power line of your phone is one of a bundle of the lines that make the space on and under your table messy.

Fig. 1. Cables messing the space under your desk in the office.

If the office is shared by more than two workers and their desks are located close to one another, the floor and your leg-rooms are populated by a lot of power lines for the computers and fax machines. Though they are the last thing to show formal visitors, we can't do without them for our work. This might have motivated people to imagine the cordless PCs and electronic devices that are still empowered by electric energy from the power outlet. Especially, considering wireless communication systems and remote controlled home appliances we use, it might not be very challenging to seek the solutions of the wireless power transmission and reception[1,2]. It is possible to transmit the electric power to a receiving electronic product using the conventional RF technologies, but it is noticed that the way of characterizing and designing a WPT system is not exactly the same as that of linking mobile devices, since the frequency ranges are different. When RF communication systems and WPT components exist together, the design will get more complicated than the purely WPT case? Here is a picture of this situation.

Customarily, the WPT design is done independent of the wireless communication, but Fig. 2 is the right picture that makes us prepared for the near future having the antennas of a laptop and LED communication circuitry. In this chapter, the electromagnetic interference between WPT and RF parts is ruled out for the sake of convenience. Prior to the explanation on the design approaches, we need to look back upon what kinds of techniques have been suggested to couple a transmitter and a receiver in a WPT system. Generally, there are magnetically coupled resonators for short distance WPT system.

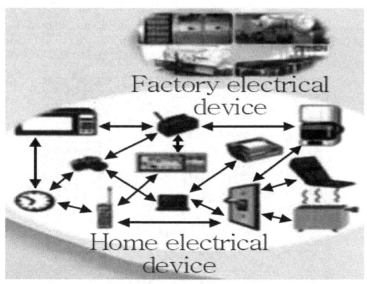

Fig. 2. WPT for home appliances and factory facility[2]

Fig. 3 is a test set-up for WPT. Seen in textbooks on circuit and Electromagnetics, the magnetic field created by the electric current of the transmitter(or loop 1) reaches the receiver(or loop 2) and induces the electric current. This is the product of so called electromotive force Faraday investigated. Actually, since the electric current is alternating current(or AC), it results in electromagnetic fields and radiation, but the frequency of the current is low and the distance between them makes magnetic field stronger than radiated fields and waves. The intensity of the magnetic induction is affected by the radius of a loop and the number of loops(or turns), but the total length of the metallic wire does not determine the frequency. And it is very reactive. However, what if the wire resonates at the frequency of operation to increase the quality factor of the energy transfer?

Fig. 3. WPT tested in a semi-anechoic chamber[2].

Fig. 4. Resonant loops are used in the university of Incheon. WPT system.

From the experience of RFID system designs, magnetic coupling has a short range, but antennas as resonators of a reader and a tag have a higher coupling value at an increased distance[3]. Magnetic resonators(electromagnetic resonance is correct) or resonant loops are installed for a 60W power transfer experiment with a 2m distance in the figure above. So this chapter describes the design approaches for the resonance type of WPT. The next section is assigned to the circuit approach which is followed by the full-wave simulation iterative design.

2. Brief introduction to the circuit theory approach for the WPT system design

The metallic loops and the field play the roles of resonators and coupling elements in a filter system. This is why the circuit theory can be adopted. The following structure is interpreted as resonators coupled each other and ports.

Using the quasi-static formulae shown in the textbooks of Electromagnetics, RFID design handbook or what not[3-5], the geometry results in the circuit elements of its equivalent circuit model that helps designers know the input impedance[6].

The details of the same approach will be revisited in other chapters by H. Hirayama and M. Mongiardo et al of this book. About Fig. 5(a), we check the coupling from the transmitter to the receiver as in Fig. 6. The resonance frequency of the WPT is 13.56MHz and the size of the receiver is less than 10cm×10cm.

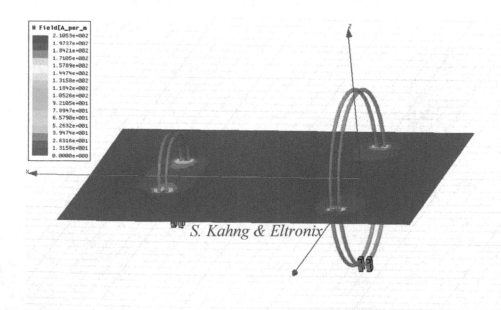

(a) Real geometry of theWPT system

Fig. 5. (Continued)

Enhanced Coupling Structures for Wireless Power Transfer Using the Circuit Approach and the Effective Medium
Constants (Metamaterials)

177

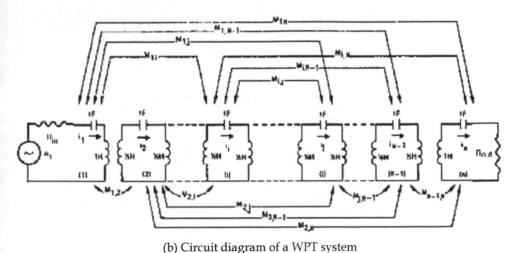

(b) Circuit diagram of a WPT system

Fig. 5. WPT system is interpreted as coupled resonators.

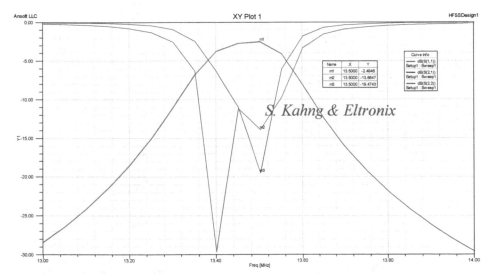

Fig. 6. An indirect-fed example obtained by the present authors.

3. RFID-antenna-inspired full-wave simulation approach and new MTM resonators

Though the near-field RFID and WPT have different purposes and frequencies, it is not too difficult to find something in common to the two areas. They use similar styles of resonance and impedance matching in the near-field zone linkage, and the same formulae of self- and mutual loop inductance. So in this section, the resonance frequency is set as 900MHz, and the full-wave solution approach adopted in the RFID antenna design is used, and a new concept metamaterial loop resoantor to increase the transfer coefficient with no change in the size is presented. Please note the full-wave solver is the FIT method where the size of a mesh is 1 tenth of wavelegth and the 8 layered PML is used as the absorbing boundary condition. First, an ordinary resonant loop is desinged(as a transmitter or a receiver).

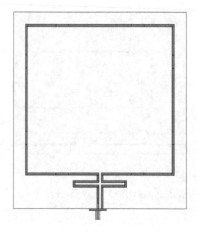

Fig. 7. A conventional rectangular resonant loop.

As this is an ordinary rectangular loop, it has multiple resonances as the harmonics above the first resonance.

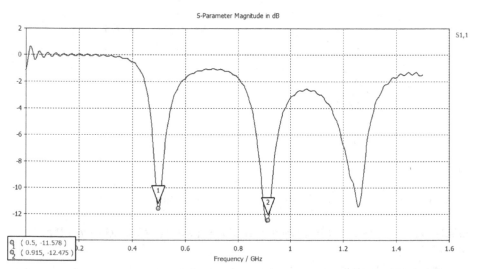

Fig. 8. Return loss(or S_{11}) of the conventional rectangular resonant loop.

The resonance at 900MHz shows the best impedance match from the figure above. So the frequency is used to couple the power source to the power load. We have plotted the he magnetic field distribution at a near-field distance over the ordinary resonant loop.

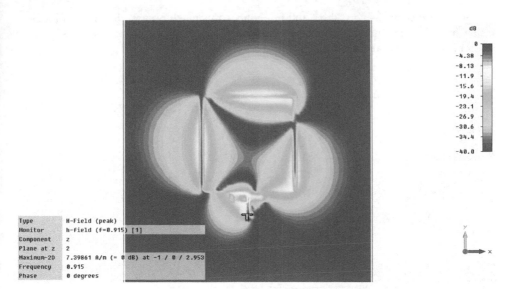

Type	H-Field (peak)
Monitor	h-field (f=0.915) [1]
Component	z
Plane at z	2
Maximum-2D	7.39861 A/m (= 0 dB) at -1 / 0 / 2.953
Frequency	0.915
Phase	0 degrees

Fig. 9. Magnetic field distribution in the vicinity of the conventional rectangular resonant loop.

Because the standing waves of half-wavelength and its interger multiples are created along the loop at resonances, we can see several null points through which the direction of electric currents changes. Due to this fact, the magnetic field has weak points fromed around the central axis of the inside of the loop. Using the loop, we can check the transfer coefficient of the WPT with different distances. The following set-up has the two ordinary resonant loops placed with a 5cm-distance.

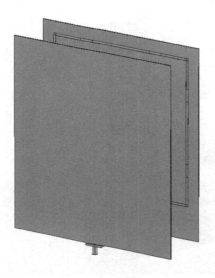

Fig. 10. The two ordinary resonant loops are 5cm away.

Though their distance is very short and impedance mismatch occurs, the original loops are neither modified nor tuned for the best condition, for the sake of convenience. Its S_{21} as the transfer coefficient is calculated as

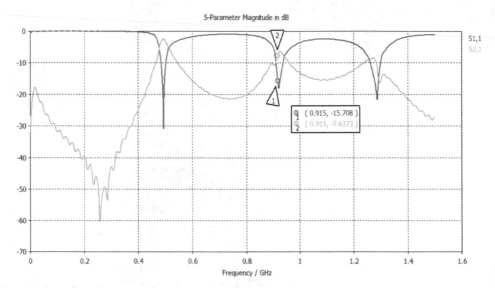

Fig. 11. S_{21} when the two ordinary resonant loops are 5cm away.

S_{21} at 900MHz is read -7dB with the unchanged loops in a changed environment. Next, the distance between the conventional rectangular resonant loops becomes 10cm. Still, the frequency of 900MHz is observed.

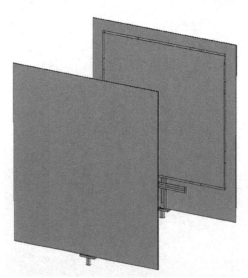

Fig. 12. The two ordinary resonant loops are 10cm away.

Though their distance is still very short and the impedance of the loop is mismatched, the original loops are neither modified nor tuned for the best condition, for convenience. Its S_{21} as the transfer coefficient is calculated as

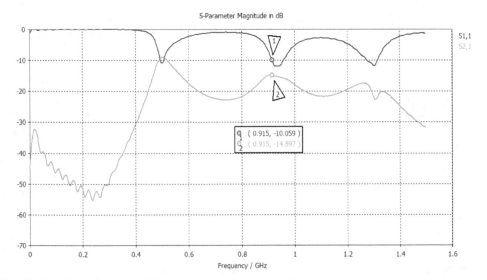

Fig. 13. S_{21} when the two ordinary resonant loops are 10cm away.

S_{21} at 900MHz is read -15dB with the unchanged loops in a changed environment. The transfer coefficient has dropped by 8dB according to the increased distance. Next, the conventional rectangular resonant loops are placed with the distance of 25cm. The frequency of 900MHz is still observed.

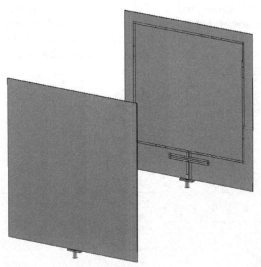

Fig. 14. The two ordinary resonant loops are 25cm away.

While their distance has increased and the impedance mismatch of the loop has got mitigated compared to the former two cases, the original loops are neither modified nor tuned for the best condition, for convenience. Its S_{21} as the transfer coefficient is calculated as

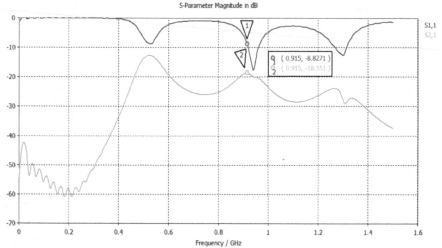

Fig. 15. S21 when the tow ordinary resonant loops are 25cm away.

S_{21} at 900MHz is read -18dB with the unchanged loops in a changed environment. The transfer coefficient has dropped by 11dB from the distance of 5cm. It is inferred that the increased distance can be a reason to experience the degraded transfer efficiency along with the impedance mismatch due to the fixed geometry of the loops and a relatively high frequency Another reason of the weakened coupling is that the ordinary resonant loop can't get the maximum value of the magnetic field along the center axis. The magnetic field in the axis determines the coupling and efficiency between the transmitter and receiver.

So by devising a metamaterial ZOR loop hinted from [7-9], we would like to create the the maximum value of the magnetic field along the center axis with no change in the size of the loop and can improve the WPT coupling.

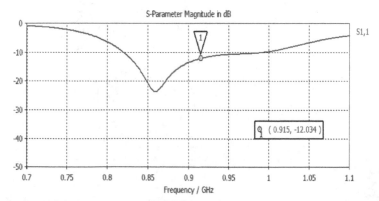

Fig. 16. Return loss of the metamaterial ZOR loop.

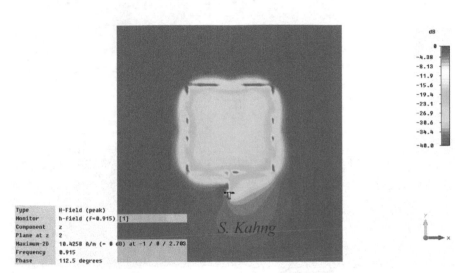

Fig. 17. Magnetic field distribution in the vicinity of the MTM ZOR loop.

The center axis passing through the loop has the maximum and almost uniform distribution of the magnetic field which will help the energy transfer enhanced. Using the new loops, we can check the transfer coefficient of the WPT with different distances. The following set-up has MTM ZOR loops placed with a 5cm-distance.

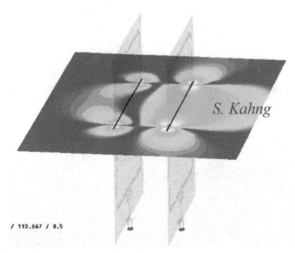

Fig. 18. The two MTM ZOR loops are 5cm away.

Though their distance is very short and serious impedance mismatch occurs, the original MTM ZOR loops are neither modified nor tuned for the best condition, for the sake of convenience. Its S_{21} as the transfer coefficient is calculated as

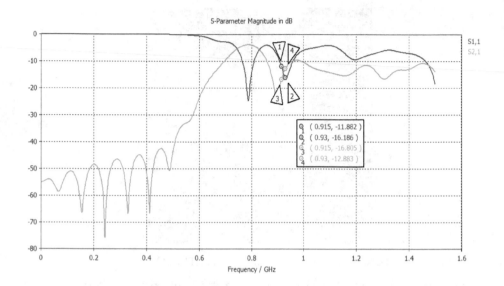

Fig. 19. S_{21} when the two MTM ZOR loops are 5cm away.

S_{21} at 900MHz reads -12dB with the unchanged loops in a changed environment.Actually, it is admitted that the MTM ZOR loop is not much superior to the conventional case for the near-field reactive ranges. Next, the distance between the MTM ZOR loops becomes 10cm.

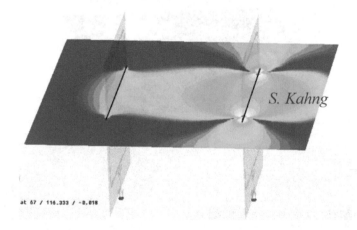

Fig. 20. The two MTM ZOR loops are 10cm away.

Still, the frequency of 900MHz is observed.

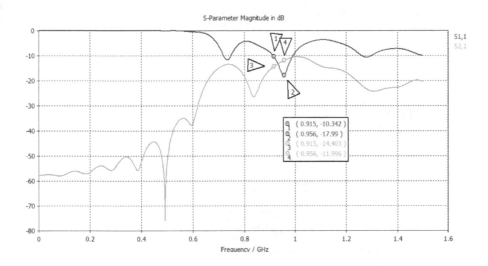

Fig. 21. S$_{21}$ when the two MTM ZOR loops are 10cm away.

S21 at 900MHz reads approximately the same as the 5cm-case with the unchanged loops in a changed environment. It is good to see the transfer coefficient keep constant with the increased distance, which can't be expected in the conventional case. Next, the MTM ZOR loops are placed with the distance of 25cm.

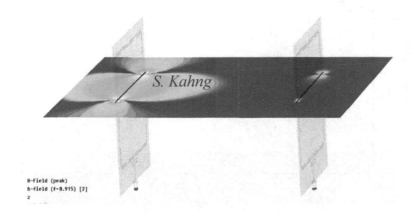

Fig. 22. S$_{21}$ when the two MTM ZOR loops are 25cm away.

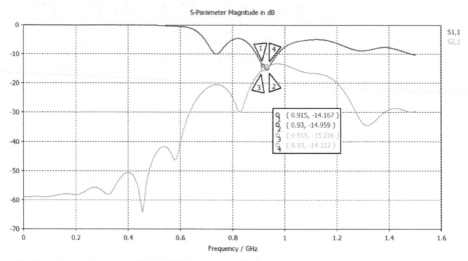

Fig. 23. S$_{21}$ when the two MTM ZOR loops are 25cm away.

S21 at 900MHz reads -14dB with the unchanged loops in a changed environment. The transfer coefficient has dropped by the increased distance, but the MTM ZOR loops have better energy transfer efficiency than the conventional design method, while they have the same area of the loop. We have investigated the advantages of the new structure and the shortcomings of the conventional loops based upon the full-wave simulation approach. This is summarized with the following comparative data.

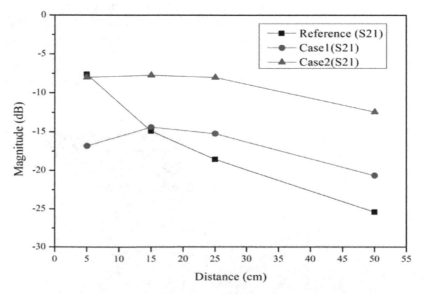

Fig. 24. Proposed technique(Case 1) and its modification(Case 2) compared to the conventional WPT.

The reference, case 1 and case 2 mean the conventional loops, the MTM ZOR loops and the reflector-backed MTM ZOR loops, respectively. My research group, maintaining the maximum magnetic field around the center axis due to the MTM ZOR, wanted to enhance the coupling and added and adjusted the reflectors to back the MTM ZOR WPT system which turns out the best in this experiment.

4. Conclusion

The necessity and background of the WPT was briefly tapped into, and two design schemes to make WPT systems were addressed. The design schemes are based on the circuit theory and the full-wave simulation. The circuit theory is used to obtain the initial and deterministic design parameters of the WPT system comprising resonators and their coupling elements and achieve a passband at the frequency of interest. The near-field RFID inspired full-wave simulation approach is to characterize EM fields of transmitter and receiver loops and their coupling, which is fed back to the design and get a right result. Going further from one method, hybridizing the two methods works well and several examples were presented. In particular, one of them is the enhanced efficiency of the WPT devising MTM ZOR loop by my research group.

5. References

[1] W.C. Brown, "The history of power transmission by radio waves," IEEE Transactions on Microwave Theory and Techniques, Vol. 32, No. 9, pp. 1230–1242, September 1984.

[2] J. H. Yoon, W. J. Byun, J. I. Choi, and H. J. Lee, "Analysis of RF Energy Transmission Technology to Realize Industry," Electronics and Telecommunication Technical Trend Analysis Report, ETRI Vol. 26, No. 4, August 2011

[3] Klaus Finkenzeller , RFID Handbook: Fundamentals and Applications in Contactless Smart Cards, Radio Frequency Identification and Near-Field Communication, Wiley, 2010

[4] Hiroshi Hirayama, Equivalent circuit and calculation of its parameters of magnetic-coupled resonant wireless power transfer, a chapter of this book

[5] Mauro Mongiardo, Alessandra Costanzo and Marco Dionigi: Networks methods for the analysis and design of wireless power transfer, a chapter of this book

[6] S. Kahng, "Study of the design of waveguide filters with improved suppression of modal interference through the cross-shaped slot," International Journal of RF and Microwave Computer-Aided Engineering, Vol. 13, no. 4, pp.285-292, 2003

[7] S. Kahng et al, "Design of a dual-band metamaterial bandpass filter using zeroth order resonance," Progress In Electromagnetics Research C, Vol. 12, 149-162, 2010

[8] S. Kahng et al, "A novel metamaterial CRLH ZOR microstrip patch antenna capacitively coupled to a rectangular ring," Antennas and Propagation Society International Symposium (APS/URSI), July, 2010

[9] S. Kahng et al, "Design of a Metamaterial Bandpass Filter Using the ZOR of a Modified Circular Mushroom Structure," Microwave Journal, Vol. 54, No.5, 158-165, 2011

10

A Fully Analytic Treatment of Resonant Inductive Coupling in the Far Field

Raymond J. Sedwick

University of Maryland, College Park, Maryland,
USA

1. Introduction

The principal behind Resonant Inductive Coupling (RIC) was first recognized and exploited by Tesla (Tesla, 1914), and its potential is often seen demonstrated by the operation of the eponymic Tesla coil. His work on RIC then, just as with the work of many groups now, was focused on the development of a means for wirelessly transmitting power. While Tesla's goal was much more ambitious (global transmission of power) the more modest goal of most current research is to power small electronics over a range of several meters. A resurgence of interest occurred in large part due to a an analysis (Karalis, 2007) where Coupled Mode Theory (CMT) was used to provide a framework to predict and assess the system performance over medium range distances. These distances are characterized as being large in comparison to the transmit and receive antennas, but small in comparison to the wavelength of the transmitted power. An evaluation of the various loss modes (to be discussed shortly) showed that for antennas made from standard conductors, the maximum power coupling efficiency occurs near 10 MHz, where the combination of resistive and radiative losses are at a minimum. The effective range of these systems – a few meters at non-negligible efficiencies – is adequate to power personal electronics (laptops, cell phones) or other equipment within a room.

Follow-on research (Sedwick, 2009) investigated the performance benefit that could be achieved by eliminating the ohmic losses of the coils through the use of high temperature superconducting (HTS) wire. It was shown that this allowed for the frequency to be lowered, with a corresponding reduction in radiative losses, providing an overall increase in efficiency over even 100's of meters. Because of the lower frequency, a higher self-capacitance of the coil could be tolerated allowing for a more compact "flat spiral" design, rather than the helical coil geometry used by Karalis. The ribbon geometry that is typical for HTS wire[1] is also well suited to forming such a flat spiral, and both the self-inductance and self-capacitance of the resulting structure (and therefore the resonant frequency) can be estimated analytically in the limit that the inter-turn spacing is small in comparison to the wire width. The analytic formulation of the resonant frequency from the geometry of the coil forms the basis of the current treatment.

A limitation to operating a superconducting version of RIC is the need to cryogenically cool the HTS components. This is more easily achieved on the transmit side of the system where

---

[1] See for instance "American Superconductor" (http://www.amsc.com/)

power is plentiful, but requires a bit of bootstrapping on the receive side, where enough power must be delivered to both supply the load and power the thermal control system. One solution is to develop a hybrid system, whereby the transmit antenna is superconducting but the receive antenna is not. The performance of such a system is expected to fall somewhere between the fully superconducting and fully non-superconducting versions, and the first part of this paper provides a model to predict this hybrid performance. The remainder of the paper then looks at the application of a hybrid RIC system to close-range communications that would be impervious to the attenuation experienced by radiative systems.

2. Extension of the performance model

The analysis of Karalis, et al. and later that of Sedwick employed CMT (Haus, 1984) as a framework for treating the system as a set of first order linear differential equations in power amplitude. Using a resonant circuit, a complex amplitude corresponding to the instantaneous power is defined

$$a = \sqrt{\frac{C}{2}}v + j\sqrt{\frac{L}{2}}i \quad W = a^*a = \frac{C}{2}V_{max}^2 = \frac{L}{2}I_{max}^2 \tag{1}$$

where C, L are the capacitance and inductance, V, I are the peak voltage and current, v, i are the instantaneous voltage and current and W is the total energy contained within the circuit. With this definition, losses that are small with respect to the recirculated power are introduced using coupling parameters as

$$\dot{a}_1 = j\omega_0 a_1 - \Gamma_1 a_1 + j\kappa_{12} a_2$$
$$\dot{a}_2 = j\omega_0 a_2 - \Gamma_2 a_2 + j\kappa_{21} a_1 - \Gamma_W a_2 \tag{2}$$

where $\Gamma_1 a_1$, $\Gamma_2 a_2$ are unrecoverable drains to the environment, $\kappa_{12} a_2$, $\kappa_{21} a_1$ are each exchanges with the other resonant device and $\Gamma_W a_2$ is delivered to the load. It can be shown by energy conservation that under this definition the coupling coefficients must be equal ($\kappa_{12} = \kappa_{21} = \kappa$). While previously it was assumed that the oscillators were identical, the possibility is considered in this work that the system is composed of inhomogeneous elements. Each coil is assumed to be of similar construction, with a ribbon wire of width w and thickness t wound into a flat spiral having N turns. The spacing between consecutive turns is d, and any dielectric in the coil (to support inter-wire spacing or to force a lower resonant frequency) will have a relative dielectric constant ε_r and a loss tangent $\tan\delta$. Depending on the design details, the losses in the coil can be ohmic (R_O), radiative (R_R), and dielectric (R_D), where each loss is expressed in terms of resistance in Ohms. The functional dependencies of these losses on the design parameters are given by

$$R_O = \rho\frac{l}{A} = \frac{4\pi^2 RN}{w}\sqrt{\rho f} \approx 3.9\frac{RN}{\overline{w}}\sqrt{\overline{\rho}\overline{f}} \qquad [R]=m, \ [\overline{w}]=mm$$

$$R_R \approx \frac{8\pi^3}{3}\sqrt{\frac{\mu_0}{\varepsilon_0}}\left(\frac{NA}{\lambda^2}\right)^2 \approx 0.2 \ (NR^2\overline{f}^2)^2 \qquad [\overline{\rho}]=\mu Ohm - cm \tag{3}$$

$$R_D = \frac{\tan\delta}{\omega C} = \omega L\tan\delta \approx 316 N^2 R\overline{f}\tan\delta \qquad [\overline{f}]=10^7 \ Hz$$

where the inductance of the spiral coil has been approximated as $L = 4\mu_0 RN^2$. The units of the quantities are chosen to be most representative of typical values that might be encountered in a design and result in numerical values of order unity. In the case of the dielectric losses, loss tangents will typically be in the 10^{-4} to 10^{-3} range, depending on the material. Designs with dielectrics in the coils can typically be avoided, however the presence of tuning capacitors (unless dielectricless) will require this loss mechanism to be included. In this case, only the portion of the capacitance that has the dielectric should appear and the expression in terms of inductance would require modification. In the present work, any capacitance in the system is assumed from the coil itself. The ohmic loss relationship assumes that the frequency is high enough that the skin depth, rather than the wire thickness (t) determines the current-carrying cross-section, which is typically the case. From Eqs. (1) and (2), the rate of energy dissipation leads to a definition for Γ.

$$\frac{dW}{dt} = -2\Gamma W = -2\Gamma\left(\frac{L}{2}I_{max}^2\right) = -\frac{I_{max}^2}{2}R_{diss} \quad \Rightarrow \quad \Gamma = \frac{R_{diss}}{2L} \tag{4}$$

and the coupling coefficient is defined in terms of the mutual inductance (M) by

$$\kappa = \frac{\omega M}{2(L_1 L_2)^{1/2}} \approx \left(290\frac{\sqrt{R_1 R_2}}{D}\right)^3 \bar{f} \tag{5}$$

where the number of turns in each coil is seen to cancel. The coils have been assumed axially aligned as in (Sedwick, 2009). The power coupled from the primary to the secondary coil depends on the relative magnitudes and phases of the energy recirculating through them. This can be seen by considering the rate at which energy is lost from the primary, as given by

$$\frac{d}{dt}|a_1|^2 = \dot{a}_1 a_1^* + a_1 \dot{a}_1^* = -2\Gamma|a_1|^2 - 2\kappa(a_1^R a_2^I - a_2^I a_1^R) \tag{6}$$

where $()^R, ()^I$ refer to the real and imaginary components of the complex amplitudes of each coil. This phase dependence leads to the familiar exchange of energy between weakly coupled systems as shown in Fig. 1 for a pair of un-driven resonant pendula.

In this figure, one pendulum (the Primary Oscillator) starts at a maximum energy and the other starts at zero energy. In Eq. (6), this is equivalent to $a_1^R = a_2^I$ having a maximum value and $a_1^I = a_2^R = 0$. It is important to recognize that the amplitudes refer to the black envelope lines in the figure, which oscillate at a frequency that is determined by the coupling constant (κ), rather than the blue curves, which oscillate at the resonant frequency of the system. The actual positions of both pendula at $t = 0$ are at their respective equilibrium locations (zero offset angle), however the velocity of the first pendulum is a maximum, whereas the velocity of the second pendulum is zero. After a quarter period, all of the energy has been transferred to the second pendulum, except for that which has been lost by dissipation (Γ). Energy then flows back to the first pendulum over the next quarter period.

Without loss of generality, we will always choose this phase relationship, and we will consider the steady-state situation where power is continuously supplied to the primary coil at the same rate that it is lost (both through dissipation and coupling to the secondary) and power is continuously coupled to the secondary coil at the same rate that it is lost (both

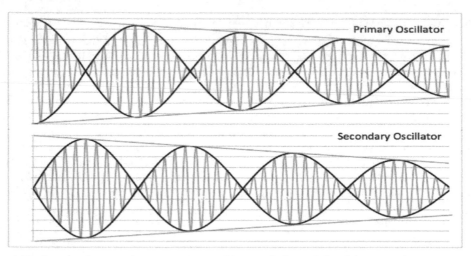

Fig. 1. Exchange of energy between two weakly coupled pendula of the same frequency.

through dissipation and driving the load). In this state, both a_1^R and a_2^I will remain constant, although not necessarily having the same value. We will also recognize that for this choice of phasing, $|a_1| = a_1^R, |a_2| = a_2^I$.
The steady-state condition for the secondary coil is therefore given by

$$\kappa |a_1||a_2| = (\Gamma_2 + \Gamma_W)|a_2|^2 \quad \Rightarrow \quad \omega M I_1 I_2 = (R_R + R_O + R_D + R_W)I_2^2 \tag{7}$$

the first form of which was used in (Karalis, 2007; Sedwick, 2009) to eliminate the amplitudes from the definition of the efficiency. The second form shows the relationship between the currents in the primary and secondary coils, which will generally be separation and orientation dependent as a result of the mutual inductance. The power available at the load is the difference between the coupled power and the power dissipated by the secondary coil, as illustrated in Fig. 2. Using Eq. (7) and the definitions of Eq. (1), the power to the load is given as

$$\frac{P_W}{I_2^2} = R_W = \frac{4\pi^3}{D^3}(N_1 R_1^2)(N_2 R_2^2)\left(\frac{I_1}{I_2}\right)\bar{f} - 0.2(N_2 R_2^2)^2 \bar{f}^4 - \left(\frac{4R_2 N_2 \sqrt{\bar{\rho}}}{\bar{w}}\right)\sqrt{\bar{f}} - 316 N_2^2 R_2 \tan\delta\bar{f} \tag{8}$$

where the power has been normalized by the square of the current in the secondary coil, resulting in an expression that is essentially the resistance of the load as seen by the secondary coil.

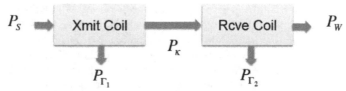

Fig. 2. Schematic of power flow from source to load showing coupling and loss paths.

Previously, an expression of this form was used to find a frequency that optimized performance, assuming that the other parameters were fixed. However, for the close-packed spiral coil (of ribbon wire) being considered here, and assuming no other reactive components are present, the natural frequency of the coil is given by (Sedwick, 2009)

$$\bar{f} \approx \frac{0.86}{R} \sqrt{\frac{\bar{d}}{\varepsilon_r N \bar{w}}} \quad \Rightarrow \quad NR^2 \bar{f}^2 \approx \frac{0.75 \, \bar{d}}{\varepsilon_r \, \bar{w}} = \beta \tag{9}$$

where $\bar{d}, \bar{w}$ appear from the calculation of coil capacitance. For other wire geometries the parameters and numerical factors that appear on the right side may change, but the product $NR^2 \bar{f}^2$ should persist, making it approximately constant from one coil to another for a given wire packing and insulation. Inserting this into Eq. (8) we find

$$\bar{I} = \frac{I_2}{I_1} = \frac{4\pi^3 \beta^2}{(\bar{f}D)^3} \left[R_W + 0.2\beta^2 + \left(\frac{4\beta}{R_2 \bar{w}} \right) \sqrt{\frac{\bar{\rho}}{\bar{f}^3}} + \frac{316\beta \tan \delta}{(R_2 \bar{f})^3} \right]^{-1} \tag{10}$$

where the number of turns in the coil has been eliminated in lieu of the coil radius where appropriate. It is interesting to note that comparison of Eqs. (8) and (9) shows that the radiation resistance is completely specified by the details of the wire geometry and insulation. For a given primary coil current, the power delivered to the load at the secondary is then

$$P_W = \frac{I_2^2}{2} R_W = \frac{I_1^2}{2} R_W \bar{I}^2 \quad \Rightarrow \quad \bar{I} \frac{\partial \bar{I}}{\partial \bar{f}} = 0 \tag{11}$$

since the currents are given as peak values rather than RMS. Also shown is the condition that would result in a frequency that maximizes the power delivered. However, in the current form there is no longer a peak in the power delivered. Instead, the power delivered increases toward lower frequencies, corresponding to an ever-increasing number of turns, since the radius has been assumed fixed in Eq. (10). At lower frequencies the effects of dielectric and ohmic losses are seen to become problematic, leading to the dielectricless, superconducting design considered in (Sedwick, 2009). The thermal overhead of a superconducting coil in some cases may be prohibitive, but a design that avoids the use of a dielectric is critical to achieving peak performance.

Differentiating the second expression for power given in Eq. (11) with respect to the load resistance (assuming I_1 is fixed), the maximum power is delivered when the load resistance is equal to the total dissipation, a result also found in (Sedwick, 2009). This can be seen to result from the fact that while a larger load will increase power for a given current, it will also decrease the amount of current through Eq. (10). The maximum power occurs when the load is as large as possible without overly limiting the current, i.e. when it is equal to the dissipation.

2.1 Two superconducting, dielectricless coils

For the superconducting case (no ohmic or dielectric losses, and $R_W = 0.2\beta^2$), the parameter β is seen to cancel out of the current ratio, making the result independent of the wire packing and producing the same equation identified as the figure of merit (FOM) given in

(Karalis, 2007; Sedwick, 2009). Using Eq. (9), the maximum power delivered in the superconducting case can be expressed solely in terms of coil design parameters as

$$P_W^R \approx I_1^2 N_2^3 \left(\frac{\overline{w}}{\overline{d}} \right) \left(\frac{6.1 R_2}{D} \right)^6 \tag{12}$$

where it is seen that only the secondary coil parameters are present. The geometry of the primary coil is of course constrained implicitly by Eq. (9), but this coil need not be superconducting.

For the power delivered in the non-superconducting case (but still with no dieletric), the resistive losses will dominate, and the load resistance should be matched to this quantity instead. In this case the expression for maximum power delivered is given as

$$P_W^O = I_1^2 N_2^{2.25} \left(\frac{\overline{w}}{\overline{d}} \right)^{1/4} \frac{\overline{d}}{\sqrt{\overline{\rho} R_2}} \left(\frac{3.4 R_2}{D} \right)^6 \tag{13}$$

The ratio of maximum power delivered by a system with only radiative losses to one with mainly ohmic losses is then

$$\frac{P_W^R}{P_W^O} \approx 33 \frac{\sqrt{\overline{\rho} R_2}}{\overline{d}} \left(\frac{N_2 \overline{w}}{\overline{d}} \right)^{3/4} \tag{14}$$

which will typically be on the order of a few thousand. For the same power delivered, this would mean a difference in range of a factor of about 3 to 4.

To find the overall system efficiency we need to evaluate the amount of power required at the input of the primary coil. This can be found in a similar way to the power to the load in terms of the coupled power and the coil losses given by

$$\frac{P_S}{I_1^2} = \frac{P_\kappa + P_{\Gamma_1}}{I_1^2} = \frac{4\pi^3 \beta^2}{(\overline{f} D)^3} \left(\frac{I_2}{I_1} \right) + 0.2\beta^2 + \left(\frac{4\beta}{R_1 \overline{w}} \right) \sqrt{\frac{\overline{\rho}}{\overline{f}^3}} + \frac{316\beta \tan \delta}{(R_1 \overline{f})^3} \tag{15}$$

where the substitutions using Eq. (9) have already been made. As before we will first assume the case when the coil is superconducting and with no dielectric, in which case the last two terms are zero. Assuming the maximum power transfer condition, the efficiency can be written

$$\eta = \frac{P_W^R}{P_S^R} = \frac{P_\kappa - P_R}{P_\kappa + P_R} = \frac{x - \alpha \overline{I}}{x + \alpha / \overline{I}} \quad \text{where} \quad x = \frac{4\pi^3 \beta^2}{(\overline{f} D)^3}, \ \overline{I} = \frac{x}{2\alpha}, \ \alpha = 0.2\beta^2 \tag{16}$$

Elimination of frequency using Eq. (9) results in

$$\eta = \left[2 + 4 \left(\frac{\alpha}{x} \right)^2 \right]^{-1} \approx \left[2 + \left(\frac{\overline{d}}{\overline{w} N_{1,2}} \right)^3 \left(\frac{D}{9.85 R_{1,2}} \right)^6 \right]^{-1} \tag{17}$$

where the number of turns and radius of either coil can be used as long as they are consistent. The radius and number of turns for either the primary or secondary coil can be used since they must be related by the frequency matching condition. The maximum efficiency is seen to be 50% because for the maximum power transfer the load power is equal to the dissipated power in the secondary coil, and in the limit of $\alpha \to 0$ the dissipated power in the primary coil becomes negligible to that in the secondary. For a nominal coil design, with the values $\overline{w}/\overline{d} = 4, N = 100$, the efficiency drops to 10% when the coils are separated by about 280 radii.

2.2 Two non-superconducting, dielectricless coils

For the non-superconducting case, the ohmic losses will dominate the coil. We will assume at first that both the primary and the secondary are non-superconducting. The only change from Eq. (17) is the definition of α, which is now the ohmic resistance, and is unique to each coil based on its radius. This gives

$$\eta = \left[2 + 4\frac{\alpha_1\alpha_2}{x^2}\right]^{-1} \qquad \alpha_{1,2} = \left(\frac{4\beta}{R_{1,2}\overline{w}}\right)\sqrt{\frac{\overline{\rho}}{\overline{f}^3}} \tag{18}$$

Again eliminating the frequency we find

$$\eta = \left[2 + \frac{\overline{\rho}}{\overline{w}\overline{d}N_2^{1.5}}\sqrt{\frac{\overline{d}}{\overline{w}}}\frac{R_2^2}{R_1}\left(\frac{D}{3.07R_2}\right)^6\right]^{-1} \tag{19}$$

which is shown expressed in terms the secondary coil, although the radius of the primary coil does appear. Swapping all of the indices will result in the same efficiency as a result of the frequency matching condition. For two of the nominal coils as described above, with the additional information of having identical radii of 0.5 m, wire width of 4 mm and formed from copper wire, the efficiency drops to 10% when the coils are separated by 20 radii. This is a distance of just over 9 meters, as compared to 140 meters for the superconducting case.

2.3 A hybrid set of dielectricless coils

The form of the efficiency given in Eq. (18) lends itself nicely to considering hybrid system designs, since the expressions for $\alpha_{1,2}$ can refer to coils with different losses. However, remarkably the efficiency is completely symmetric with respect to which coil is the primary versus the secondary. The reason that this is unexpected is that one of the coils could have substantially more current, and one might expect that making this coil superconducting would result in a more efficient design. Let us assume that a system has a superconducting primary and a non-superconducting secondary, and that the ratio of losses is a factor of $\lambda > 1$. Swapping coils will decrease the resistance of the secondary by λ (assuming the load remains matched to the secondary losses), resulting in an increase in $\overline{I}$ by this same factor. Assuming the same power to the load, the current in the secondary must increase by a factor of $\sqrt{\lambda}$, which means that the current in the primary must be reduced by this same factor to keep the coupled power constant. The resistance of the primary is now larger by a factor of λ (the resistive coil), so the power dissipated in the primary remains the same. If the load power and dissipated powers all remain the same, so does the efficiency.

The advantage of this is that a superconducting coil can be introduced on either side of the system and will have the same effect of increasing the system efficiency. Since the superconducting coil will have some thermal management overhead that will require power (a power which has not been accounted) it would be more useful to put it at the end of the system that is transmitting the power, where power is plentiful. However, in the case of using the system for communications (which is discussed next), both sources would have their own power supply and signals could be transmitted bi-directionally. If one of the systems has an ample supply of power, whereas the other has a limited supply of power, the superconducting coil can be placed at the power rich side for the benefit of both ends. As a final consideration in this section, the efficiency of a hybrid system is given as

$$\eta = \left[2 + 4\frac{\alpha_1\alpha_2}{x^2}\right]^{-1} \approx \left[2 + \left(\frac{\overline{d}}{\overline{w}}\right)^{1.25}\frac{\sqrt{\overline{\rho}R_2}}{\overline{w}N_2^{2.25}}\left(\frac{D}{5.50R_2}\right)^6\right]^{-1} \tag{20}$$

For a hybrid system with the parameters specified above, the efficiency drops to 10% at a distance of 75 radii, which is the geometric mean of the distances for the other two cases as might be expected. Again the efficiency is expressed in terms of the secondary coil parameters, but can be expressed in terms of the primary by changing all of the indices.

3. Application to communications

While the efficiency drops off rapidly, one application that does not require an efficient transfer of power is communications. For example, if a receiver requires a microwatt of power to get a signal, but the transmitter is operating at only one Watt of power, the efficiency is 10^{-6}, but the system works as desired. Thus far, coupled mode analysis has been used to assess the power transfer and efficiency between resonant coils of similar designs with potentially different loss characteristics. However, the only frequency present was the resonant frequency of the system. To assess the applicability of such hybrid systems for communications, an approach must be used to capture the effect of broad-spectrum coupling.

For the purpose of transmitting signals, only three coil components will be assumed: A drive coil which is non-resonant, and a pair of resonant coils, which as previously may each have different loss characteristics. Demodulation of the signal on the secondary side will be assumed to occur directly off the secondary coil since it will provide the largest voltage. The drive coil is driven at the resonant frequency of the tuned coils, but only couples strongly to the primary coil due to both its proximity and limited magnetic flux generation. The signal into the drive coil is encoded using amplitude modulated. To proceed with the analysis, a sign convention will be established whereby current traveling clockwise around a coil (as viewed from the primary to the secondary coil, assuming that they are axially aligned) is considered positive, while a counter-clockwise current is negative.

The model for all three coils is the same (see Fig. 3), with capacitive and inductive elements having series resistors to capture the various loss modes. Beginning with the secondary coil, the governing equation is given as

$$L_S\dot{I}_S + R_S I_S + \frac{1}{C_S}q_S = emf = -M_{PS}\dot{I}_P \tag{21}$$

where I_S, q_S are the current and unbalanced charge in the coil, L_S, R_S and C_S are the effective inductance, resistance and capacitance of the coil and *emf* is the electromotive

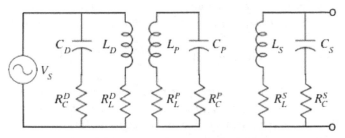

Fig. 3. Electrical model of drive, primary and secondary coils.

force induced in the secondary coil by the changing current in the primary. The resistance includes both R_C^S (dielectric loss) and R_L^S (ohmic and radiative coil losses). An increasing positive current in the primary will induce a negative *emf* in the secondary by Lentz's Law, proportional to the mutual inductance (M_{PS}) between the two coils. Differentiating with respect to time yields

$$\ddot{I}_S + 2\Gamma_S \dot{I}_S + \omega_0^2 I_S = -\frac{M_{PS}}{L_S}\ddot{I}_P \tag{22}$$

where Γ_S is the decay rate as defined previously using CMT. Similarly, the governing equation for the primary is

$$\ddot{I}_P + 2\Gamma_P \dot{I}_P + \omega_0^2 I_P = -\frac{M_{PD}}{L_P}\ddot{I}_D \tag{23}$$

where the back *emf* from the secondary has been neglected because it will be very small compared to the drive coil *emf* .

3.1 Drive coil
Treatment of the drive coil is different, since it is not driven inductively, but instead directly by an amplifier stage. It consists nominally of a single loop of wire in close proximity to the primary and has a resonant frequency that would typically be much higher than the operating frequency of the system. Current driven around the coil from the amplifier is effectively traveling through the inductive and capacitive legs in parallel, so its impedance can be found as

$$Z_D = \left(\frac{1}{R_L^D + j\omega L_D} + \frac{1}{R_C^D - j/\omega C_D}\right) = \frac{(R_L^D R_C^D + X_D^2) + jX_D(\frac{\omega}{\omega_D}R_C^D - \frac{\omega_D}{\omega}R_L^D)}{(R_L^D + R_C^D) + jX_D(\frac{\omega}{\omega_D} - \frac{\omega_D}{\omega})} \tag{24}$$

where the resistance values are as shown in Fig. 3, and X_D is the reactance of both the inductor and capacitor at the drive coil's resonant frequency (ω_D). Since this frequency is so much higher than the operating frequency of the primary and secondary coils, the impedance of the drive coil reduces to

$$Z_D = R_L^D + j\omega L_D \tag{25}$$

at all relevant frequencies. For a given applied voltage, $V_D(\omega)$, the current in the coil is simply found from Ohm's law. Assume an amplitude modulated voltage of the form

$$V_D = V_m \sin \omega_0 t \, (1 + \gamma \cos \omega t) = V_m \, \text{Im}[\bar{V}_D], \quad \bar{V}_D = e^{j\omega_0 t} + \frac{\gamma}{2}(e^{j(\omega_0 + \omega)t} + e^{j(\omega_0 - \omega)t}) \quad (26)$$

where V_m is the amplitude of the carrier, γ is the amplitude of the modulation relative to the carrier, ω_0 is the carrier frequency (resonant frequency of the primary and secondary coils) and ω is the frequency of the modulation. The complex form will be used throughout the analysis for simplicity, allowing the appropriately phased solution to be found at any point by taking the imaginary component.

It should be noted that a substantial back *emf* from the primary will be seen across the drive coil, so that the voltage driving the current in the drive coil will actually be

$$V_S = V_m \bar{V}_D - M_{PD} \dot{I}_P \quad (27)$$

Knowing this, we will assume for simplicity that this back *emf* has been compensated by the amplifier, so that Eq. (26) represents V_S, the net voltage seen across the drive coil.

3.2 Primary coil
The current in the primary coil is found by solving Eq. (23), driven by $V_S \, (\sim V_D)$

$$\ddot{I}_P + 2\Gamma_P \dot{I}_P + \omega_0^2 I_P = -\frac{M_{PD} V_m}{L_P(R_D + j\omega L_D)} \ddot{\bar{V}}_D \quad (28)$$

where

$$\ddot{\bar{V}}_D = -\omega_0^2 e^{j\omega_0 t} + \frac{\gamma}{2}[-(\omega_0 + \omega)^2 e^{j(\omega_0 + \omega)t} - (\omega_0 - \omega)^2 e^{j(\omega_0 - \omega)t}] \approx -\omega_0^2 \bar{V}_D \quad (29)$$

In the last form of Eq. (29) (far right), the impact of ω versus ω_0 on the amplitude of the signal has been neglected. This is a good approximation since the modulation frequency will be small in comparison to the resonant frequency and this approximation will be made throughout the analysis. Similarly, for the current of the drive coil we will assume

$$\frac{e^{j\omega_0 t}}{R_D + j\omega_0 L_D} \approx \frac{e^{j\omega_0 t}}{j\omega_0 L_D} \quad \text{and} \quad \frac{e^{j(\omega_0 \pm \omega)t}}{R_D + j(\omega_0 \pm \omega)L_D} \approx \frac{e^{j(\omega_0 \pm \omega)t}}{j\omega_0 L_D} \quad (30)$$

where we have neglected the small relative phase shift introduced by the modulation and by the resistance of the drive coil. While the transient homogeneous solution to Eq. (28) will play a role under a continuously varying input signal, for the present analysis we are merely concerned with the frequency response of the driven system. Assuming a particular solution of the form $I_P e^{j\omega t}$, Eq. (28) becomes

$$(\omega_0^2 + 2\Gamma_P j\omega - \omega^2)I_P = -j\left(\frac{M_{PD} V_m \omega_0}{L_P L_D}\right)\bar{V}_S \quad (31)$$

Solving Eq. (31), each of the terms from $\bar{V}_S$ couple into the current of the primary as

$$\frac{e^{j\omega_0 t}}{\omega_0^2 - \omega_0^2 + 2\Gamma_p j\omega_0} \Rightarrow \frac{-je^{j\omega_0 t}}{2\omega_0\Gamma_p}, \quad \frac{e^{j(\omega_0 \pm \omega)t}}{\omega_0^2 - (\omega_0 \pm \omega)^2 + 2\Gamma_p j(\omega_0 \pm \omega)} \Rightarrow \frac{-j(\Gamma_p \mp j\omega)e^{j(\omega_0 \pm \omega)t}}{2\omega_0(\Gamma_p^2 + \omega_0^2)} \tag{32}$$

and the complete expression for the current in the primary coil becomes

$$I_p = -\frac{M_{PD}V_m}{L_P L_D 2\Gamma_p}\left[e^{j\omega_0 t} + \frac{\gamma}{2}\left(c_p^* e^{j(\omega_0 + \omega)t} + c_p e^{j(\omega_0 - \omega)t}\right)\right], \quad c_p = \frac{\Gamma_p^2}{\Gamma_p^2 + \omega_0^2} + j\frac{\omega\Gamma_p}{\Gamma_p^2 + \omega_0^2} \tag{33}$$

Because of the close proximity of the drive coil to the primary, if we assume that the radii of both coils are the same, then nearly all of the flux generated by one coil will thread the other. This is equivalent to saying that the coupling coefficient (κ) is nearly unity. However in general, the mutual coupling between the drive and primary coils can be given in terms of the inductance of either coil as $M_{PD} = \kappa\sqrt{L_D L_P} = \kappa\frac{N_P}{N_D}L_D = \kappa\frac{N_D}{N_P}L_P$. Using the first expression with $N_D = 1$, and substituting in for the physical definition of Γ_P, the primary current is

$$I_p = -\frac{\kappa N_P V_m}{(R_L^P + R_C^P)}\left[e^{j\omega_0 t} + \frac{\gamma}{2}\left(c_p^* e^{j(\omega_0 + \omega)t} + c_p e^{j(\omega_0 - \omega)t}\right)\right] \tag{34}$$

Although it may not be surprising to see the familiar Ohm's law relationship with a transformer step-up, there are two subtleties worth noting. The first is that the reactive component of the coil impedance does not appear, which is at first striking because the high Q of these coils implies that the reactive part should dominate. The second point to note is that the primary coil is not simply being driven as the "secondary" of a standard transformer, but rather at the resonance frequency of the primary. Therefore the peak voltage across the primary should well exceed the step-up resulting from simply taking the turns ratio of the transformer. These two points actually resolve themselves if we recognize that it is the reactance of the coil that actually drives it to high voltages at resonance. So, given the current found in Eq. (34), the peak voltage across the primary is not in fact $N_P V_m$, but rather $V_{peak} = I_P X_0 = I_P L_p \omega_0$. The real power dissipated in the coil is however just $I_P N_P V_m$ or

$$P_{Loss}^P = \frac{(N_P V_m)^2}{R_L^P + R_C^P} = I_P^2(R_L^P + R_C^P) \tag{35}$$

While it is true that for a given drive voltage, decreasing the primary coil resistance actually increases the power dissipation, the goal is to achieve a certain level of current. Given I_P, reducing the coil resistance reduces both the power loss as well as the voltage that must be generated by the amplifier.

We can now consider the back *emf* in the drive coil that must be compensated. This is given by

$$V_D^{emf} = -M_{PD}\dot{i}_p = -\left(\kappa\frac{N_D}{N_P}L_P\right)\frac{N_P V_m j\omega_0}{(R_L^P + R_C^P)} = -j\kappa V_m\frac{L_P\omega_0}{(R_L^P + R_C^P)} = -j\kappa V_m Q_P \tag{36}$$

so that whatever the desired V_m, the back *emf* to be compensated will lag by 90° and be greater in amplitude by a factor of κQ_P, the product of the coupling constant and the quality of the primary coil.

Looking now at the complex part of the primary current, taking the imaginary component results in

$$I_P = -\frac{\kappa N_P V_m}{(R_L^P + R_C^P)} \sin \omega_0 t \left[1 + \gamma \left(\frac{\Gamma_P^2}{\Gamma_P^2 + \omega^2} \cos \omega t + \frac{\Gamma_P \omega}{\Gamma_P^2 + \omega^2} \sin \omega t \right) \right] \quad (37)$$

and in the desired low-loss limit, this becomes approximately

$$I_P = -\frac{\kappa N_P V_m}{(R_L^P + R_C^P)} \sin \omega_0 t \left[1 + \gamma \frac{\Gamma_P}{\omega} \sin \omega t \right] \quad (38)$$

where low-loss assumes that the dissipation factor is small in comparison to the modulation frequency. If we consider at what frequency the modulated signal is reduced to 50% of its original value, we find

$$\omega = 2\Gamma_P \quad \Rightarrow \quad \frac{\omega}{\omega_0} = \frac{2}{\omega_0} \frac{R_P}{2L_P} = \frac{R_P}{L_P \omega_0} = \frac{1}{Q_P} \quad (39)$$

So, as expected, the bandwidth that can be supported by the system is governed by the resonant frequency and the quality of the oscillator. To achieve both high efficiency and bandwidth, the frequency of operation should then be as high as possible. Since high frequency corresponds to radiative losses in an inductively coupled system of a given size, this sets fundamental constraints on efficiency and bandwidth. However, it also indicates that reducing the ohmic and dielectric losses in favor of radiative losses is a way to maximize bandwidth and efficiency. If it is desired to maintain low detectability, the radiated energy must be shielded in some way such as by using a Faraday cage.

3.3 Secondary coil

Continuing now to the secondary coil, the governing equation for the current is given by Eq. (22), with the driving current given by Eq. (34). As before, $\ddot{I}_S \approx -\omega_0^2 I_S$ resulting in

$$(\omega_0^2 + 2\Gamma_S j\omega - \omega^2) I_S = \frac{M_{PS}}{L_S} \omega_0^2 I_P = \frac{M_{PS}}{L_S} \frac{N_P V_m \omega_0^2}{(R_L^P + R_C^P)} \left[e^{j\omega_0 t} + \frac{\gamma}{2} \left(c_P^* e^{j(\omega_0 + \omega)t} + c_P e^{j(\omega_0 - \omega)t} \right) \right] \quad (40)$$

After similar machinations as before, the current in the secondary can be given as

$$I_S = \frac{M_{PS}}{L_S} \frac{N_P V_m \omega_0^2}{(R_L^P + R_C^P)} je^{j\omega_0 t} \left[1 + \frac{\gamma}{2} \left(c_S^* c_P^* e^{j\omega t} + c_P^* c_P e^{-j\omega t} \right) \right], \quad c_S = \frac{\Gamma_S^2}{\Gamma_S^2 + \omega_0^2} + j \frac{\omega \Gamma_S}{\Gamma_S^2 + \omega_0^2} \quad (41)$$

Expressing the mutual inductance and self-inductances in terms of system parameters as in Section 2, the amplitude of the current in the secondary becomes

$$|I_S| = \frac{M_{PS} N_P V_m \omega_0}{2(R_L^P + R_C^P)\Gamma_S L_S} = \left(\frac{N_P^2 R_P^2}{(R_L^P + R_C^P)} \right) \left(\frac{N_S^2 R_S^2}{(R_L^S + R_C^S)} \right) \frac{4\pi^3 V_m \overline{f}}{D^3} \quad (42)$$

Dividing out the current in the primary we find

$$\frac{I_2}{I_1} = \left(\frac{N_P R_P^2 N_S R_S^2}{(R_L^S + R_C^S)}\right)\frac{4\pi^3 \overline{f}}{D^3} = \left(\frac{N_P R_P^2 N_S R_S^2}{R_W + R_{Rad}}\right)\frac{4\pi^3 \overline{f}}{D^3} =$$

$$= \left(\frac{\beta^2 / \overline{f}^4}{R_W + 0.2\beta^2}\right)\frac{4\pi^3 \overline{f}}{D^3} = \frac{4\pi^3 \beta^2}{(\overline{f}D)^3}\left[R_W + 0.2\beta^2\right]^{-1} \tag{43}$$

where we have eliminated all losses except for radiative, added in an equivalent series resistance for the load and substituted in definitions from Eqs. (3) and (9). Gratifyingly, the final expression on the right is identical to that of Eq. (10), under the same set of assumptions. Taking only the imaginary component of Eq. (41), the current corresponding to our assumed input in the low-loss operating limit is

$$I_S = \left(\frac{N_P^2 R_P^2}{(R_L^P + R_C^P)}\right)\left(\frac{N_S R_S^2}{(R_L^S + R_C^S)}\right)\frac{4\pi^3 V_m \overline{f}}{D^3}\cos\omega_0 t\left[1 - \gamma\frac{\Gamma_P \Gamma_S}{\omega^2}\cos\omega t\right] \tag{44}$$

Considering the coefficient of the modulated waveform as before, the bandwidth based on a 50% reduction in amplitude is given by

$$\frac{\Gamma_P \Gamma_S}{\omega^2} = \frac{1}{4 Q_P Q_S}\left(\frac{\omega_0}{\omega}\right)^2 = \frac{1}{2} \quad\Rightarrow\quad \Delta f = \frac{f_0}{\sqrt{2 Q_P Q_S}} \tag{45}$$

and can be seen to be related to the geometric mean of the primary and secondary coil Q-factors.

As an example, we will assume an operating frequency of 10 MHz, a frequency where the radiative losses begin to be comparable to ohmic losses. Voice transmission has a bandwidth typically taken to be 4kHz[2], so the combined system Q-factor can be as high as 1800 using Eq. (45). If radiative losses are the only dissipation in the system, then the Q-factor of each coil can be calculated as

$$Q = \frac{L\omega_0}{R_{Rad}} = \frac{32\pi^2 N^2 R\overline{f}}{0.2(NR^2\overline{f}^2)^2} = \frac{160\pi^2}{(R\overline{f})^3} \quad\Rightarrow\quad R\overline{f} = \frac{11.6}{\sqrt[3]{Q}} \tag{46}$$

At 10 MHz, $\overline{f} = 1$, so $R = 0.95m$ from Eq. (46), but the number of turns in the coils is unspecified since both radiative resistance and reactance scale as N^2. At 1MHz, the Q-factor would need to be 10 times smaller to support the required bandwidth, requiring the coil radius to be over 21 times larger. However, there is no need to require that the dissipation be purely radiative, and only a small amount of additional dissipation is needed to substantially reduce the Q-factor.

4. Range of detection and bandwidth

For a given demodulation threshold, the ability to detect and demodulate the signal from the carrier is distance dependent by the overall strength of the received signal. The

---

[2] See for instance http://en.wikipedia.org/wiki/Voice_frequency

maximum voltage that will be seen across the secondary coil can be found from the current induced and the Q-factor of the coil, as was discussed with the primary coil. The modulated signal will then be attenuated as shown in Eq. (45). For the case of a superconducting secondary, these can be combined to yield

$$V_S^{max} = I_S Q_S = I_P \left(\frac{I_S}{I_P}\right)\left(\frac{1}{4Q_P Q_S}\right)\left(\frac{\omega_0}{\omega}\right)^2 Q_S = \frac{I_P}{Q_P}\left(\frac{5\pi^3}{\overline{f}^3 D^3}\right)\left(\frac{10^7 \overline{f}}{\Delta f}\right)^2 \approx \frac{I_P}{Q_P \overline{f}(\Delta f^*)^2}\left(\frac{2490}{D}\right)^3 \quad (47)$$

where Q_P and I_P have been left explicitly and the bandwidth (Δf^*) is in kHz. The reason is that all of the factors in the final form would be operationally derived. For instance, while an arbitrarily high primary current may be theoretically possible, power limitations or wire current capacity would most likely limit it to a maximum value. Likewise, there is no constraint on making the Q-factor of the primary large or small, so we allow it to be specified directly. The Q-factor of the secondary is seen to cancel, since while it reduces the current in the secondary it also increases the peak voltage. Although this cancellation occurs (making its value irrelevant), the relationship used for the current ratio has already assumed that only radiative losses are seen in the secondary and that the load power is negligible by comparison. In general, however, Eq. (10) can be used to find I_S / I_P for any dominant loss mode.

The achievable bandwidth and carrier frequency are not independent of the values of Q_P and Q_S, and Eq. (45) can be used to conservatively replace the carrier frequency in lieu of the Q-factors. This results in

$$V_S^{max} \approx \frac{(10^4) I_P}{Q_P \sqrt{2 Q_P Q_S}(\Delta f^*)^3}\left(\frac{2490}{D}\right)^3 \quad \Rightarrow \quad \Delta f^* \approx \left(\frac{I_P}{V_S^{max} Q_P \sqrt{Q_P Q_S}}\right)^{\frac{1}{3}}\left(\frac{47800}{D}\right) \quad (48)$$

Under the present assumptions, a selection of threshold voltage, primary coil current, Q-values, and operating distance will determine the bandwidth. As an example, assume that the detection threshold is 20 mV, about four times the typical laboratory noise floor. Parallel windings of wire could allow for a recirculating primary current of 100 Amps, and quality voice communications would require $\Delta f = 4kHz$ $\left(\Delta f^* = 4\right)$. High Q-factors are seen to be detrimental, however since the secondary has been assumed superconducting, the expectation is that Q_S will nevertheless be quite large – on the order of 10^6. The variation of bandwidth with distance for three values of Q_P is shown in Fig. 4.

A consequence of using a high-Q coil as a secondary is that the Q-factor of the primary must be smaller to support a given bandwidth. This translates into higher power requirements on the primary. For the case of a receiver coil that is dominated by ohmic losses, the modulated signal strength across the secondary coil is given instead by

$$V_S^{max} = I_P \left(\frac{I_S}{I_P}\right)\left(\frac{1}{4Q_P Q_S}\right)\left(\frac{\omega_0}{\omega}\right)^2 Q_S = \frac{I_P}{Q_P}\frac{\pi^3 \beta R_2 \overline{w}}{D^3 \sqrt{\overline{f}^3 \rho}}\left(\frac{10^7 \overline{f}}{\Delta f}\right)^2 \approx \frac{I_P \beta R_2 \overline{w}}{Q_P (\Delta f^*)^2}\sqrt{\frac{\overline{f}}{\rho}}\left(\frac{1460}{D}\right)^3 \quad (49)$$

where the ohmic term in Eq. (10) has been used for the current ratio between the secondary and the primary. Using the same substitution to eliminate the carrier frequency results in

$$V_S^{max} \approx \frac{(Q_P Q_S)^{1/4}}{Q_P \sqrt{\bar{\rho}}} \frac{I_P \beta R_S \bar{\omega}}{(\Delta f^*)^{3/2}} \left(\frac{353}{D}\right)^3 \quad \Rightarrow \quad \Delta f^* \approx \left(\frac{(Q_P Q_S)^{1/4}}{Q_P \sqrt{\bar{\rho}}} \frac{I_P R_S \bar{d}}{V_S^{max}}\right)^{\frac{2}{3}} \left(\frac{321}{D}\right)^2 \tag{50}$$

The frequency dependency now weakly favors a higher Q_S, which is now unfortunately much smaller as a result of the ohmic losses in the secondary. Using the additional assumptions of $R_S = 0.5m$, $\bar{d} = 1mm$, $\bar{\rho} = 1.6$ along with the same previously assumed parameters, the bandwidth is plotted as a function of distance in Fig. 5 for the same values of Q_P assumed in Fig 4.

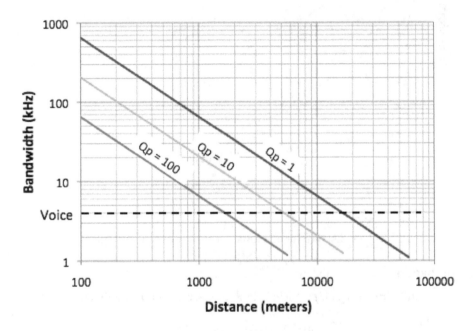

Fig. 4. Bandwidth versus distance for a superconducting secondary at different qualities.

The range is seen to be smaller overall for the ohmic loss case, and in both cases the distance is seen to increase as the Q-factor of the primary is decreased. The implication of reducing the primary Q-factor is a greater amount of power that must be expended. It should be noted that by fixing the Q-factors of the primary and secondary coils for the performance curves shown in Figs. 4 and 5, the assumed carrier frequency is changing for different values of bandwidth. These plots, therefore, do not show the reduction in bandwidth with distance of a fixed system design, but rather how the baseline design would change for different maximum ranges of operation. To see how the bandwidth of a particular system would roll off with distance would require going back to Eqs. (47) and (49).

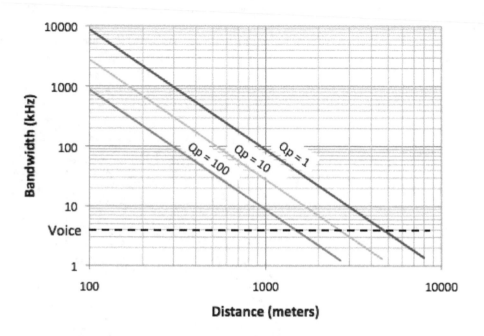

Fig. 5. Bandwidth versus distance for an ohmic secondary at different qualities.

5. Modulation

Writing the full form of Eq. (47) to include both the carrier and the modulation we find

$$V_S = I_P \left(\frac{20\pi^3}{f^3 D^3} \right) \cos \omega_0 t \left[1 - \frac{\gamma}{4} \frac{1}{Q_P Q_S} \left(\frac{\omega_0}{\omega} \right)^2 \cos \omega t \right] \tag{51}$$

so that in the current example, the amplitude of the modulated signal in comparison to the DC component has a numerical value of only 8(10⁻⁵). This may make the demodulation difficult due to small variations in the carrier that could be confused with signal content.

The assumed modulation approach thus far has been a Double Sideband Full Carrier scheme (DSFC—standard AM), which has the advantage of using a simple envelope detector for demodulation. The envelope detector circuit and the resulting idealized envelope are shown in Fig. 6. Instead of transmitting a full carrier along with the signal (the unity term in Eq. (48), setting γ to a large value effectively eliminates the DC component of the modulation. This transforms the signal to a Double Sideband Suppressed Carrier (DSB-

SC) AM signal. Elimination of the carrier reduces the power required for transmission and removes variations in the carrier as a potentially competing signal. The disadvantage is that without the carrier present the signal must be mixed at the receiver with a new carrier signal that is phase-locked and frequency matched to the original. Standard methods of implementing such a system are available.

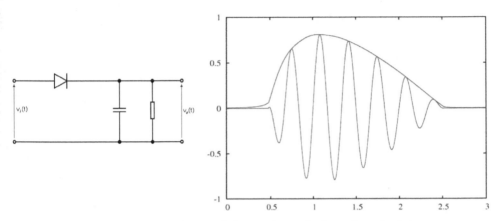

Fig. 6. Envelope detector circuit (left) and the resulting ideal envelope (right).

For communications the bandwidth is seen to roll-off with frequency for different point designs, and this reduction is demonstrated to be linear for a superconducting secondary and quadratic for a non-superconducting secondary. A large Q-factor secondary has the advantage of operating over larger distances, however the consequential need for reducing the Q-factor of the primary causes the power dissipation in the primary to be larger. Conversely, a non-superconducting secondary suffers a reduced range of operation, but levies a lower power requirement on the primary coil as a result. A system can be optimized to meet a specific bandwidth/distance requirement with the lowest power consumption.

6. Acknowledgments

The author wishes to acknowledge the support of the Defense Advanced Research Projects Agency (DARPA) Strategic Technologies Office (STO) monitored by Dr. Deborah Furey under PO No. HR0011-10-P-003. The author would also like to acknowledge Dr. Eric Prechtl of Axis Engineering Technologies, Inc. for his technical contributions to the effort.
The views expressed are those of the author and do not reflect the official policy or position of the Department of Defense or the U.S. Government. This work has been approved for Public Release, Distribution Unlimited.

7. References

Haus, H. A., Waves and Fields in Optoelectronics, Prentice-Hall, New Jersey (1984)
        http://en.wikipedia.org/wiki/Voice_frequency
Karalis, A., et al., "Efficient wireless non-radiative mid-range energy transfer", Ann. Phys., 10.1016 (2007)

Sedwick, R.J.,"Long range inductive power transfer with superconducting oscillators", Ann. Phys., 325 (2010) 287–299.

Tesla, N., U.S. patent 1,119,732 (1914)

11

AC Processing Controllers for IPT Systems

Hunter Hanzhuo Wu[1], Grant Anthony Covic[2] and John Talbot Boys[2]

[1]Energy Dynamics Laboratory, Utah State University Research Foundation
[2]Department of Electrical and Computer Engineering, The University of Auckland
[1]USA
[2]New Zealand

1. Introduction

Inductive Power Transfer (IPT) systems allow electrical energy to be transferred over a relatively large air gap via high frequency magnetic fields. Such systems can be broadly classified based on application, but all such systems have various key components that include; the power electronic transmitter, the receiver electronics, the magnetic coupler and the power flow controller. The purpose of using a power flow controller is to regulate the power delivered to a load using a reference independent of system parameter variations of the IPT system. Some of the common parameters which can cause variations in the power transfer of an IPT system are load resistance and coupling coefficient. Recently, a new type of power controller called an AC processing pickup has been proposed with significant advantages in terms of increasing system efficiency, reducing pickup size and lowering production cost compared to traditional pickups that also produce a controlled AC output using complex AC-DC-AC conversion circuits. This new controller is in two forms but both forms regulate AC power directly and produce a controllable high-frequency AC source over a wide load range suitable for lighting and EV charging applications. This chapter outlines the details and operation of the two different types of AC processing pickups along with analytic analysis. Simplified design examples with practical implementation methods will be presented for systems with variations in load and coupling.

2. Parallel AC processing pickups

A parallel AC processing pickup capable of producing an AC controllable current source is described below.

2.1 Fundamentals of the AC processing pickup

A schematic circuit for a parallel AC processing pickup coupled to a resonant IPT track is shown in Fig. 1. The primary track is assumed to be energised using a resonant supply which operates at fixed frequency and adjusted to keep the track current I_1 nominally constant independent of load. Capacitor C_2 is tuned to pickup inductor L_2 at the frequency of this primary track current I_1 to form a resonant tank. The diodes (D_1, D_2) and switches (S_1, S_2) form an AC switch. From standard IPT theory, a pickup coil placed in the vicinity of the primary track would have an open circuit voltage (V_{oc}) source induced within it that is given by:

$$V_{oc} = j\omega M I_1 \tag{1}$$

where ω is the operating angular frequency, M is the mutual inductance and I_1 is the primary track current.

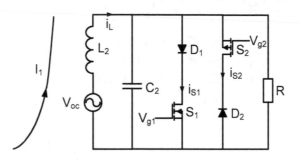

Fig. 1. The parallel AC Processing Pickup.

To illustrate the circuit operation, Fig. 2 shows the waveforms when the switches are used to clamp parts of the resonant capacitor voltage. V_{g1} and V_{g2} are the PWM gate signals used to drive switches S_1 and S_2 at 50% duty cycle with the same switching frequency as the IPT track frequency. Consider the situation where waveforms V_{g1} and V_{g2} are controlled with a phase delay ϕ relative to the phase of V_{oc} as shown in Fig. 2. At time $t=0$, S_1 is turned off and S_2 is turned on. However, the series diode D_2 blocks any current flowing through S_2 as it is reverse biased. This causes the capacitor C_2 to resonate with the pickup inductance L_2 like a parallel resonant tank. The capacitor voltage rings to a peak value and returns back to zero.

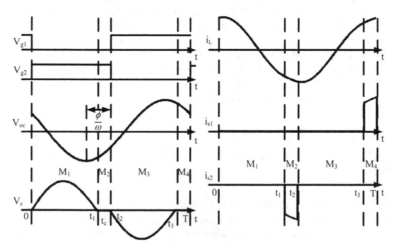

Fig. 2. Typical Single Cycle Operating Waveforms when the Switches Clamp the Resonant Capacitor Voltage for the Parallel AC Processing Pickup.

When the capacitor voltage reaches zero, D_2 terminates the resonant cycle and prevents the capacitor voltage building up in the negative direction as it begins to conduct at zero volts, thereby clamping the output voltage to zero. This causes S_2 to clamp V_c for a time known as the clamp time (t_c) at the point where V_c changes from a positive to a negative voltage. In summary, the clamping action from the AC switch generates a phase shift between the open circuit voltage and the capacitor voltage waveform. A single cycle operation for the AC processing pickup is composed of a sequence of linear circuit stages with each corresponding to a particular switching interval as illustrated to Fig. 3. These can be grouped into the following modes:

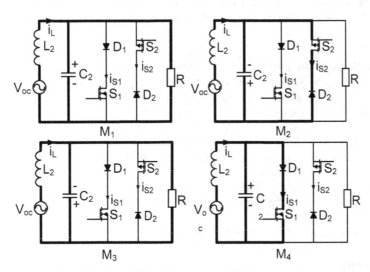

Fig. 3. Operating Modes of the Parallel AC Processing Pickup.

Mode 1 (M_1): At $t=0$, S_1 is turned off and S_2 is turned on. The capacitor voltage V_c will resonate with the pickup inductor to a positive peak voltage and then decay. Since D_2 is reversed biased due to positive V_c, it blocks any current that flows through S_2. Under this mode, the circuit operates like a parallel resonant tank and current flows into the load resistor.

Mode 2 (M_2): At $t=t_1$, the capacitor voltage naturally crosses zero. Since S_2 is still turned on, D_2 becomes forward biased, the resonant cycle is terminated and the capacitor voltage is clamped. In this mode, the inductor current flows through switch S_2 and no current flows through the load.

Mode 3 (M_3): At $t=t_2$, S_1 is turned on and S_2 is turned off. Similar to M_1, the circuit operates like a parallel resonant tank and current flows into the load resistor.

Mode 4 (M_4): At $t=t_3$, the capacitor voltage naturally crosses zero. Similar to M_3, the resonant cycle is terminated and the capacitor voltage is clamped. In this mode, the inductor current flows through switch S_1 and no current flows through the load. After this mode, the circuit returns back to M_1, and the switching process is repeated.

The AC processing pickup achieves approximate soft switching conditions. From Fig. 3, the resonant inductor current starts to flow through S_2 at t_1 when there is no voltage across it, hence ZVS is achieved at turn on. When S_2 is turned off at t_2, the resonant capacitor in parallel with S_2 forces the voltage across S_2 to increase slowly in the negative direction while the current through it decreases to zero. For most practical switches, the turn off is much faster than the rate of increase of the capacitor voltage, so the dv/dt across the switch is relatively small and ZVS is obtained at the switch off condition. Switch S_1 operates in a similar manner and also achieves ZVS at turn on, while achieving a low dv/dt at turn off. Likewise, for diodes D_1 and D_2, low dv/dt is achieved at turn on, and ZVS is achieved at turn off. In summary, the switches and diodes in the AC processing pickup achieve soft switching. This gives the pickup desirable characteristics such as low switching losses, low switching stress and reduced electromagnetic interference (EMI) levels. The low switching losses gives this pickup controller a very high operating efficiency. Moreover, the low EMI provides little interference on the control circuitry of the pickup and external systems nearby.

2.2 Exact analysis of the parallel AC processing pickup

It has been shown in (Wu et al., 2009a) that the AC processing controller regulates the output power by controlling phase delay. In this section, an exact analysis in the time domain is proposed to determine the characteristics of the circuit under steady state operation. The basis of the analysis method is that the conditions existing in the circuit at the end of a particular switching period must be the initial conditions for the start of the next switching period, and these conditions must be identical, allowing for changes in polarity caused by the resonant operation.

The analysis procedure is simplified using the same assumptions:

1. The Equivalent Series Resistance (ESR) of both capacitor C_2 and Inductor L_2 are very small and can be neglected.
2. The switching action of the transistors and diodes are instantaneous and lossless.

2.2.1 Basic equations

Let the tuning capacitance of the circuit be,

$$C_2 = aC_{20} \tag{2}$$

where

$$C_{20} = 1/(\omega^2 L_2) \quad a \in |R| \tag{3}$$

With reference to Fig. 4, the waveform can be separated into two operating modes known as the resonant mode and the clamp mode.

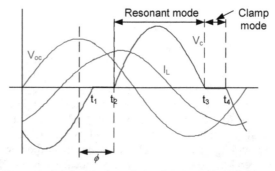

Fig. 4. Waveform in Two Modes for the Parallel AC Processing Pickup.

2.2.2 Resonant mode

During the resonant mode, the capacitor voltage may be described as:

$$\frac{d^2V_c}{dt^2} + \frac{1}{RC_2}\frac{dV_c}{dt} + \frac{V_c}{L_2C_2} = \frac{V_{oc}}{L_2C_2}\sin(\omega t + \phi) \tag{4}$$

Considering the initial condition $V_c(t)|_{t=0} = 0$ and $\frac{dV_c}{dt}|_{t=0} = \frac{-i_L(0)}{C_2}$, the complete solution

of the above equation is:

$$V_c(t) = V_{cd}(t) + V_{cu}(t) \tag{5}$$

where

$$V_{cd}(t) = \frac{\partial_2}{\sin(\gamma_v)} e^{-\sigma t}\sin(\omega_f t - \theta_v) \tag{6}$$

$$V_{cu}(t) = (1 - a)\partial_1\sin(\omega t + \phi) - \frac{\partial_1}{Q_{20}}\cos(\omega t + \phi) \tag{7}$$

$$Q_{20} = R_2 / (\omega L_2) \tag{8}$$

$$\sigma = 1 / 2aR_2C_{20} \tag{9}$$

$$\omega_f = \omega\sqrt{1/a - 1/(4a^2Q_{20}{}^2)} \tag{10}$$

$$\partial_1 = V_{oc} / ((1 - a)^2 + (1/Q_{20})^2) \tag{11}$$

$$\partial_2 = \partial_1((1 - a)\sin(\phi) - \cos(\phi)/Q_{20}) \tag{12}$$

$$\partial_3 = \partial_1((1 - a)\cos(\phi) + \sin(\phi)/Q_{20}) \tag{13}$$

$$\gamma_v = \tan^{-1}\left(\frac{-\omega_f\partial_2}{i_L(0)/Q_{20} + \sigma\partial_2 + \omega\partial_3}\right) \tag{14}$$

In a similar way, considering the initial condition $i_L(t)|_{t=0} = i_L(0)$ and $\frac{di_L}{dt}|_{t=0} = \frac{-V_{oc}\sin\phi}{L_2}$,

the complete solution to the inductor current is:

$$i_L(t) = i_{Ld}(t) + i_{Lu}(t) \tag{15}$$

where

$$i_{Ld}(t) = \frac{-i_L(0) - \beta_2}{\sin(\gamma_i)} e^{-\sigma t}\sin(\omega_f t - \theta_i) \tag{16}$$

$$i_{Lu}(t) = \frac{-\partial_1}{R_2}\sin(\omega t + \phi) + \frac{\partial_1 Q_{20}}{R_2}(-a(1-a) + \frac{1}{Q_{20}^2})\cos(\omega t + \phi) \tag{17}$$

$$\beta_2 = \frac{\partial_1}{R_2}\left(\sin(\phi) - Q_{20}\cos(\phi)(-a(1-a) + \frac{1}{Q_{20}^2})\right) \tag{18}$$

$$\gamma_i = \tan^{-1}\left(\frac{-\omega_f(i_L(0) + \beta_2)}{-V_{oc}\sin(\phi)/L_2 + \omega\beta_3 + \sigma(i_L(0) + \beta_2)}\right) \tag{19}$$

$$\beta_3 = \frac{\partial_1}{R_2}\left(\cos(\phi) + Q_{20}\sin(\phi)(-a(1-a) + \frac{1}{Q_{20}^2})\right) \tag{20}$$

To investigate how long the circuit stays in the resonant mode, $V_c=0$ can be substituted in (5), resulting in the following expression:

$$V_c(t_r) = 0 \tag{21}$$

where t_r is the time the circuit operates in the resonant mode.

2.2.3 Clamp mode
During the clamp mode, the inductor L_2 is shorted and the current depends on V_{oc}. By Kirchhoff's Voltage Law, the inductor current equation can now be written as:

$$i_L(t) = -\frac{V_{oc}}{L}\int_{t_r}^{t}\sin(\omega t + \phi)dt + i_L(t_r) \tag{22}$$

Solving (22), the inductor current can be expressed as:

$$i_L(t) = (V_{oc}\cos(\omega t + \phi))/(\omega L_2) + i_L(t_r) - (V_{oc}\cos(\omega t_r + \phi))/(\omega L_2) \tag{23}$$

Because both the resonant mode and the clamp mode are repeated each half cycle (with only a polarity change), the relationship $i_L(0) = -i_L(T/2)$ must hold. Hence, the capacitor voltage and inductor current must be given by,

$$V_c(t) = \begin{cases} V_{cd}(t) + V_{cu}(t) & t_0 \le t < t_1 \\ 0 & t_1 \le t < t_2 \\ -V_{cd}(t) - V_{cu}(t) & t_2 \le t < t_3 \\ 0 & t_3 \le t < t_4 \end{cases} \tag{24}$$

$$i_L(t) = \begin{cases} i_{Ld}(t) + i_{Lu}(t) & t_0 \le t < t_1 \\ (V_{oc}\cos(\omega t + \phi))/(\omega L_2) + i_L^- & t_1 \le t < t_2 \\ -i_{Ld}(t) - i_{Lu}(t) & t_2 \le t < t_3 \\ -(V_{oc}\cos(\omega t + \phi))/(\omega L_2) - i_L^- & t_3 \le t < t_4 \end{cases} \tag{25}$$

2.2.4 Computation routine

The above analysis is very difficult to solve analytically as the solution of t_r and $i_L(0)$ are governed by (24) and (25) with γ_v and γ_i as interim variables which are associated with the auxiliary equations (14) and (19). This is in the form of transcendental equations and it can only be solved using numeric solvers such as MATLAB or EXCEL. A computer program based on an iterative computation, shown in Fig. 5, has been developed to undertake the analysis. The program starts by initializing the circuit parameters such as L_2, C_2 and R_2. The initial condition of the inductor current is first set to the solution given by (5) and (15) in the routine as an educated guess. With $i_L(0)$ known, t_r can be calculated solving (21) and the inductor current when the resonant mode ends can be calculated using (15). With t_r known, the inductor current after half a period can be calculated using (23). The next step is to check whether $i_L(0)$ and $-i_L(T/2)$ have converged to within a given error ($\varepsilon < 0.01\%$). If the answer is YES, the program terminates and the correct solution is deemed to be found. Otherwise the iteration repeats itself in the computation loop until a solution is found. The algorithm has proven to be both fast and robust.

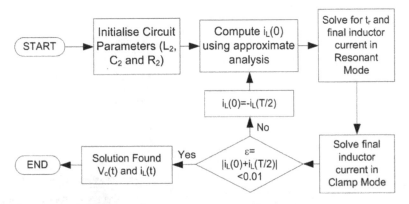

Fig. 5. A Flow Chart of Computation Algorithm.

2.2.5 Pickup characteristics

Using the exact analysis method proposed from section 2.2.1 to 2.2.3, the normalized pickup characteristics are computed. The output voltage (or capacitor voltage) characteristics of the pickup are shown in Fig. 6(a) for different values of Q_{20} or load condition. The normalized output voltage is defined by the ratio of the output voltage over the open circuit voltage. It can be seen that the output power asymptotically decreases as the controlled phase delay ϕ increases from zero. It should be noted that the output voltage can be controlled from the maximum permissible value of the parallel resonant tank ($Q_{20}V_{oc}$) down to zero. The normalized output current is shown in Fig. 6(b) for a range of Q_{20} values. Fig. 6(b) shows that the output current stays approximately constant as the load resistance changes for pickups at high Q_{20} (5-10). As such, the parallel AC processing pickup demonstrates controllable current source behaviour.

The output current-voltage characteristic is shown in Fig. 7(a). The current source behaviour is again demonstrated as the output current stays approximately constant for a given phase delay irrespective of output voltage as long as the output voltage is reasonably high.

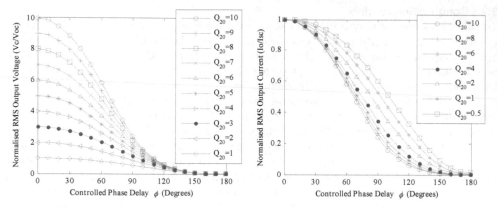

Fig. 6. (a) Normalized RMS Output Voltage vs. Controlled Phase Delay ϕ, (b) Normalized RMS Output Current vs. Controlled Phase Delay ϕ.

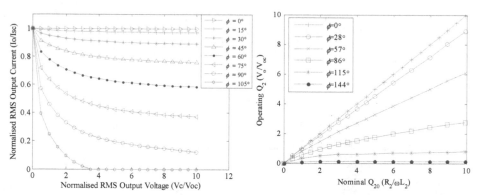

Fig. 7. (a) Pickup Output Voltage Current Characteristics, (b) Operating Q_2 vs. Nominal Q_{20}.

A plot of the secondary circuit's operating Q_2 versus its nominal operating Q_{20} is shown Fig. 7(b). The current source behaviour is again demonstrated as the slope of output voltage stays constant as the nominal Q_{20} increases for a given phase delay.

The normalized reflected resistance and reactance are shown in Fig. 8(a) and Fig. 8(b), respectively. The normalization is performed with respect to Z_t given by:

$$Z_t = \frac{\omega M^2}{L_2} \tag{26}$$

It has been found that the maximum impedance (including both its resistance and reactance) does not reach the peak when the converter is delivering full power at low Q_{20} (Wu et al., 2010). The maximum reflected impedance is actually 40% higher than the value of the Q_{20} of the circuit. This shows that if the secondary pickup requires power proportional to Q_{20}, the power supply has to source more than the real required power. However, for pickups with a Q_{20} above 3 it was found that the overrating of the normalized impedance of the primary converter is less than 10% (Wu et al., 2010). Most commercial systems use IPT pickups with the highest Q_{20} obtainable (Q_{20}=3-10) to minimise the primary track current required. Under these conditions, the overrating required by the power supply is insignificant.

Fig. 9(a) shows the transient response of the AC processing pickup with a step in the controlled phase ϕ from 20-50° at time equal to 0.5ms. Similarly, Fig. 9(b) shows the transient response of the AC processing pickup with a step in the controlled phase ϕ from 50-20° at time equal to 0.5ms. Simulation results using a fundamental mode analysis discussed in (Wu et al., 2010) are also added to the same figure for comparison purposes. Since such a simplified transient analysis only models the fundamental component, the peaks of the capacitor voltage calculated using (Wu et al., 2010) will be slightly less than the simulation results of the actual circuit. Nonetheless, this first order approximation can be used to provide a good rule of thumb for the capacitor voltage under transient operation. It can be seen that the transient response of the pickup is always over damped with no overshoot or undershoot, so the components in the pickup only have to be rated for steady state operation.

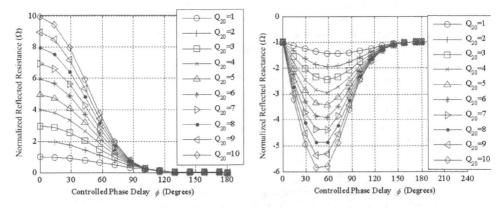

Fig. 8. (a) Reflected Resistance on Primary Track, (b) Reflected Reactance on Primary Track.

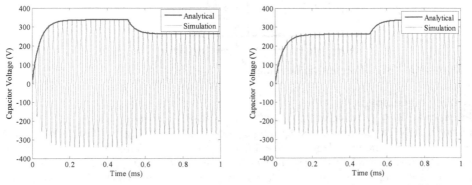

Fig. 9. (a) Capacitor Voltage when Controlled Phase Delay ϕ Steps from 20-50°, (b) Capacitor Voltage when Controlled Phase Delay ϕ Steps from 50-20°.

2.3 Lighting applications using the parallel AC processing pickup

Stage lighting systems have a tedious setup procedure involving many cables and plugs all connecting to the inverters below the scaffold. In addition such stage lights are usually constrained as to the position where they can be placed and any movement is severely limited due to the cables. The contactless nature of IPT enables IPT powered stage lights to be placed anywhere along a suitably installed track. Moreover, such a system can provide power to moving trolleys that could be used to carry lights, thereby providing extra flexibility for the stage lighting system.

Presently, most IPT systems produce a controlled DC output for industrial loads. In these applications, the most common secondary pickup controller rectifies the AC power, which is then regulated using a DC shorting (decoupling) switch. Such a pickup controller has the advantage of simple control circuitry and the ability to operate over a wide load range (Boys et al. 2000). The detailed operation of the pickup controller can be found in (Boys et al., 2000). In order to power high frequency AC loads such as fluorescent lights or stage lights, an extra resonant converter or DC-AC PWM inverter is required to produce a controllable AC source. One method of achieving this is to simply cascade the IPT pickup controller with a push-pull converter as shown in Fig. 10. However, the addition of a second converter is not ideal because of the large number of components required which increase cost. In addition, the two stage conversion process has losses in each stage which reduce efficiency. Likewise, a DC-AC PWM inverter also has a high component count and even more switching losses than the resonant converter due to a hard switching operation at these high frequencies.

In this section, the use of an AC processing pickup controller to produce a controllable AC power source over a wide resistive load range is presented. It uses a single conversion stage and improves the efficiency, economics and weight of the system. The pickup is used as a light dimmer to control the power delivered to a stage light while achieving high frequency soft switching conditions, contactless energy transmission, electric isolation and high efficiency.

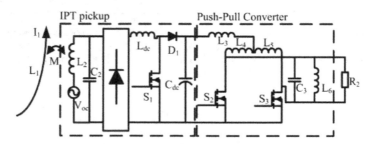

Fig. 10. An AC-DC-AC Conversion Topology.

2.3.1 Circuit description

A design of a 1.2kW AC processing pickup used in stage lighting applications is outlined in this section. A block diagram of the lighting system is shown in Fig. 11. The switch current detection method is shown in Fig. 12. When the circuit enters the clamp state, the body diodes immediately start to conduct the resonant pickup current. Detecting this change allows the start of the clamp state to be precisely determined. A current transformer (CT) can be used to detect the switch current and transform this into a voltage signal with lower amplitude. A comparator is used to generate a 50% duty cycle square wave which may be used for the microcontroller to show when the circuit entered the clamped state as shown in Fig. 12. Because this technique measures the high current through the switch, it is less prone to noise. The RMS lamp voltage is first rectified (as shown in Fig. 11) and then filtered to DC inside the microcontroller. Next, the microcontroller samples this DC voltage using an ADC and adjusts the phase of the gate drive waveforms for the AC switch accordingly. The design specifications and system parameters are shown in Table 1. A primary LCL power supply is used operating at 20kHz and producing a track current of 125A. An asymmetrical shaped S pickup (Elliot et al, 2006) is used with an open circuit voltage of 84.9V and a short circuit current of 5.86A. The tuning capacitor is chosen lower than the nominal value of (3) with an a value of 0.9 (using (2)) for the reasons outlined in (Wu, et al., 2009b). The equivalent resistance of the Phillips Broadway 1.2kW 220V lamp changes from 3-40Ω over the full output current range. This is because the halogen bulb resistance is temperature dependent and the current flowing through it directly determines its temperature.

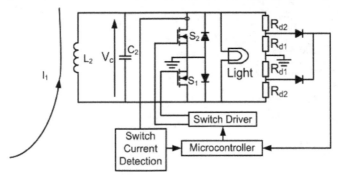

Fig. 11. Block Diagram of 1.2kW Lighting System.

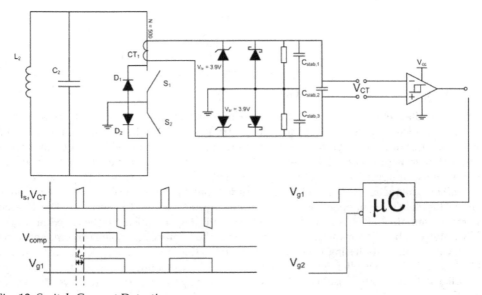

Fig. 12. Switch Current Detection.

Specification Parameters	Values	Design Parameters	Values
I_1	125A	L_2	115.2uH
P_{out}	1.2kW	V_{oc}	84.9V
$(V_c)_{rms}$	220V	I_{sc}	5.86A
R_2	$3\Omega - 40\Omega$	a	0.9
f	20kHz	C_2	495nF

Table 1. Specifications and System Parameters.

For the lighting system, clamp time is used as a key parameter to control the phase delay, which then controls the output power (Wu et al., 2011a). From Fig. 13, it can be seen that the majority of the output voltage range can be achieved by using a clamp time range of 2.7-15μs. A clamp time that is lower than 2.7μs will result in higher output voltages than the rated voltage of the lamp and these characteristics are not measured.

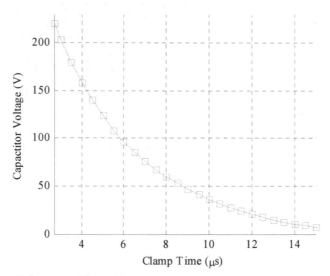

Fig. 13. Output Voltage vs. Clamp Time.

A more detailed block diagram of the microcontroller is shown in Fig. 14. As shown, a PSOC29466 microprocessor is used for the controller, as it has both analogue and digital blocks that are suited to real time control at the frequencies considered here. A second order low-pass filter with a cut off frequency of 2kHz is used to remove the high frequency components of the rectified output voltage. The cut off frequency should be chosen as high as possible to allow fast response of the system without compromising the filtering action to remove the 40kHz component and its harmonics produced by the rectifier. The filtered output is sampled at 3.2kilo-samples per second (ksps) which is the maximum the PSOC29466 can handle for this application. This is then compared against the reference set by the user. Note that aliasing is not a big concern as the input to the ADC is slow varying DC. A simple PI controller is implemented with "anti-windup". The output of the PI controller will drive the phase modulation block which generates the gate drive signals for the AC switch.

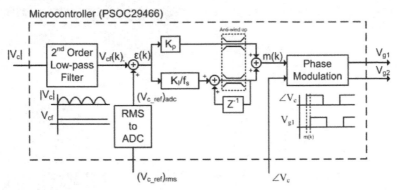

Fig. 14. Block Diagram of Microcontroller.

2.3.2 Experimental results

Fig. 15 shows the circuit waveforms when the output voltage is regulated to 220V, 140V and 50V respectively. These correspond to 100%, 50% and 10% of rated power. The waveforms show the resonant capacitor voltages or output voltages (upper trace) and the pickup inductor currents (lower trace).

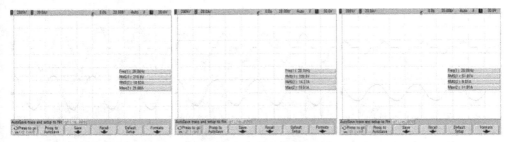

Fig. 15. Operating Waveforms of Lighting System for (a) 100%, (b) 50% and (c) 10% Output Power.

The transient response of the lighting system was quantified to be a simple first order system in (Wu et al., 2011a). Fig. 16(a) shows the step response of the output voltage from 30V – 220V. The top trace represents the lamp voltage and the bottom trace represents the lamp current. The controller is set with a PI gain of $K_p=1$, $K_I=375$ and an anti-windup limit of 2.7us clamp time. It can be seen that the rise time for the output current is 178ms and the rise time for the output voltage is 668ms. Because the bulb is a non-linear resistor that changes with temperature, the output voltage and current response times are not the same. Fig. 16(b) shows the fall time for the output current and voltage, and the response times are 424ms and 570ms.

The output current step response is largely dominated by the parallel AC processing pickup controller because of its controlled current source property. The slow response time for the output voltage is caused by the low dynamic resistance (or dR/dt) of the bulb which significantly slows down the transient response. The low dynamic resistance of the bulb adds an extra 490ms (or more than 200%) to the response time. In order to overcome this issue, more current can be driven into the light so that it heats up more quickly. This can be

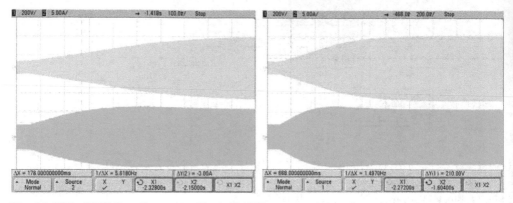

Fig. 16. 30V - 220V Step Response, K_p=1, K_I=375, tc_min=2.7us showing Rise Time of (a) Bulb Current and (b) Bulb Voltage.

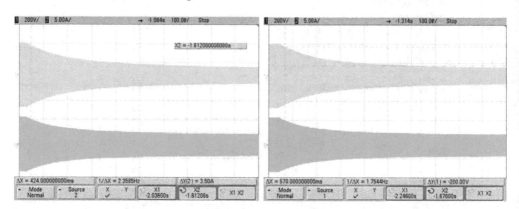

Fig. 17. 220V - 30V Step Response, K_p=1, K_I=375, tc_min=2.7us showing Fall Time of (a) Bulb Current and (b) Bulb Voltage.

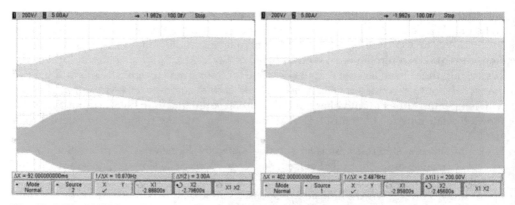

Fig. 18. 30V - 220V Step Response, K_p=5, K_I=750, tc_min=1.5us showing Rise Time of (a) Bulb Current and (b) Bulb Voltage.

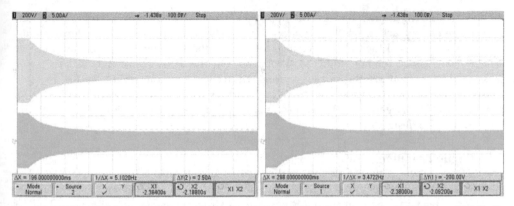

Fig. 19. 220V - 30V Step Response, K_p=5, K_I=750, tc_min=1.5us showing Fall Time of (a) Bulb Current and (b) Bulb Voltage.

achieved by reducing the minimum clamp time to 1.5us which increases the output current during the transient start up (beyond the maximum allowed steady state value shown in Fig. 13). In addition, PI gains can be increased for the pickup to yield a faster transient response. Here the K_p and K_I gains were increased to 5 and 750, respectively. Fig. 18 shows a step response of this system. It can be seen that both the current rise time and voltage rise time are much shorter – 92ms and 402ms, respectively. Fig. 19 shows the step response from 220V – 30V. The fall times for both the current and voltage are also shorter as expected. However, even if the gain is increased significantly, the output voltage response does not have a considerable change as it is dominated by the low dynamic resistance of the bulb.

2.4 Electric vehicle charging application using the parallel AC processing pickup

An AC processing pickup with cascaded rectifier is also of interest in order to produce a controlled DC output. Although such a circuit performs a similar task (producing a controllable DC) to that of a traditional pickup that utilizes decoupling (or shorting) control (Boys et al., 2000) - it has several advantages. If a decoupling controller is used in applications with large variations in coupling, an inefficient method of overrating circuit devices is required. Because the coupling can change over a wide range, the short circuit current also changes significantly. Components like the DC inductor, the shorting switch and the rectifier have to be rated for the highest short circuit current condition while normally operating at lower currents. Moreover, since the short circuit current of the parallel resonant tank is constantly flowing through the switch, this causes constant on state losses in the switch which lowers efficiency. This is particularly evident in systems whose coupling coefficient can change by over 100% (Wang et al., 2005). To design such a system, the minimum coupling coefficient must be assumed in order to ensure that the minimum current is enough to satisfy the power transfer required. However, under maximum coupling conditions the short circuit current may be as much as 100% larger than that required under normal operation, which means all of the components have to be chosen such that they will operate safely at 200% of "rated" current. This inevitably reduces efficiency and increases cost. Although primary side control could be used to lower the track current and therefore the coupled voltage, feedback signals such as output voltage and current of the secondary pickup are then required to be sent to the primary converter via a wireless communication channel. This approach is suited to applications with only one pickup, given any action on the primary side affects all coupled receivers, however the system operation will then also inevitably depend on the reliability of the communication system which is a common source of problems. On the other hand, the AC processing pickup directly controls the output current from the resonant circuit to the rectifier, consequently, no overrating is required in applications where coupling varies. This chapter proposes an IPT charger using the AC processing pickup and provides design guidelines for this new pickup design.

2.4.1 Circuit operation
The AC processing pickup described earlier can be directly cascaded with a rectifier. An alternative circuit topology that is more practical or has fewer components is shown in Fig. 20. The rectification is performed in this new circuit structure using back to back MOSFET's forming the AC switch. This has the advantage of eliminating the need for two additional diodes in the rectifier. In addition, this new circuit has a common ground reference on both the AC and DC side. This makes measurement of any voltage signal on the AC or DC side easier as the measurement circuitry does not need to be isolated. The DC inductor and capacitor acts as a filter network to produce the DC output.

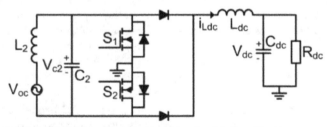

Fig. 20. An Parallel AC Processing Pickup Combined with Rectifier.

A design example for a 1kW AC processing pickup is presented in Fig. 21. Here a PSOC microcontroller is again used to measure the output voltage and adjust the phase of the switch gate drive waveforms accordingly. Some key specifications are listed on the left side of Table 2 and the designed component values are presented on the right. The pickup regulates the output voltage V_{dc} to a nominal 200V when the load varies from 40Ω to open circuit. In addition, the coupling coefficient can change by more than 100% when the height between the primary and secondary pads changes from 45mm to 90mm which is the nominal operating range for the chosen pad (Budhia et al., 2009). The primary and secondary IPT pad has a circular structure shown in Fig. 22. Dimensions of both pads are exactly identical and each pad has 8 pieces of ferrite with 12 turns. The variation in coupling with respect to height is shown in Fig. 23(a). It can be seen that the coupling factor changes from 0.16 – 0.37 as the distance between them varies from 45mm - 90mm. In addition, the variation in the inductance of the pads is shown in Fig. 23(b). The inductance of the pickup is 72.4µH and its value changes by 2.7% as the pad alignment changes. The designed a value was calculated to be 0.87 in (Wu et al., 2011b), however the exact capacitance value of 206.5nF was not available. To be conservative, an a value of 0.81 was chosen which gives a capacitance value of 192nF. The minimum V_{oc}, calculated using the equation in (Wu et al., 2011b), is 77.12V and the measured value is 78V when k=0.16. As discussed in (Wu et al.,

2011b), the DC inductance is chosen to be 10 times higher than the minimum DC inductance (L_{dcmin}) while still maintaining appropriate switch ratings.

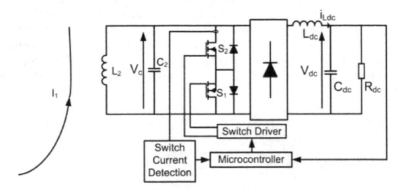

Fig. 21. Block Diagram of the AC Processing Pickup for EV Charger.

Spec. Parameters	Values	Design Parameters	Values
k	0.16 – 0.37	$(V_{oc})_{min}$	78V
P_{out}	1kW	$(I_{sc})_{min}$	4.5A
V_{dc}	200V	a	0.87
R_{dc}	40Ω - ∞Ω	g_L	1.03
I_1	29.2A	g_c	0.99
F	38.4kHz	C_2	192nF
L_0	72.4uH	L_{dc}	540uF
C_0	237.3nF	C_{dc}	1mF
L_{2e}/L_0	0.02695		
C_{2e}/C_0	0.1		

Table 2. Steady State Design Parameters for EV charging system.

All dimensions in mm

A 30   Ferrite Width
B 118   Ferrite Length
C 10   Ring Thickness
D 238   Coil Diameter
E 420   Pad Diameter
Pad Thickness 25
Coil: 12 Turns

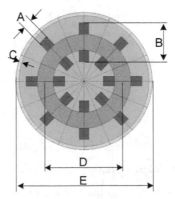

Fig. 22. EV Charger Pad Layout.

For the practical system, the output voltage vs. clamp time is shown in Fig. 24. Measurements in blue show the output voltages when operating at a distance d=90mm with a coupling coefficient k=0.16. Similarly, the measurements in red show the output voltages when operating at a coupling coefficient k=0.37 with an air gap distance d=45mm. As shown, the majority of the output voltage range can be achieved using a clamp time range of between 1-6 μs. Since the theoretical control range for this 38.4 kHz IPT system would be 0-13us, this control range is feasible and not too sensitive. To regulate the output voltage at 200V, the controller is required to measure the output voltage and vary the clamp time accordingly.

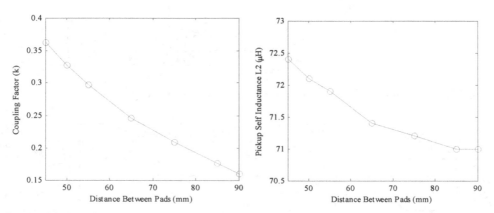

Fig. 23. (a) Coupling Factor vs. Distance Between Pads and (b) Pickup Inductance vs. Distance Between Pads.

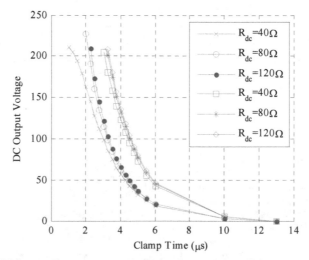

Fig. 24. Output Voltage Characteristics for Open Loop Controller.

A closed loop system was built based on the block diagram shown in Fig. 25. The output voltage is regulated to a desired reference using a simple PI controller to adjust the clamp time as shown in Fig. 25. The reference voltage can be set by a PC using serial communication. The capacitor voltage phase is measured using a detection block within the chosen microprocessor and is computed when the current starts to flow through the AC switch. A phase modulator block within the PSOC controls the phase between the output gate drive signals with respect to the phase of the capacitor voltage. Ideally this system provides a constant output voltage irrespective of coupling and load conditions. Since the change in pad inductance is inherently linked with coupling, only coupling variations are of interest as they include inductance variations.

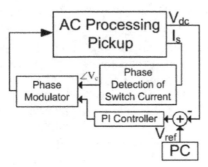

Fig. 25. EV Charging Controller Block Diagram.

Fig. 26 shows the circuit waveforms when the output voltage is regulated to 200V at k=0.16 (or d=90mm) for rated, 50% and 10% load, respectively. The waveforms are the resonant capacitor voltage, inductor current, DC inductor current and synchronization signal from the switch current detection block. Note that, the switch current detection inverts the logic which is inverted again at the output of the gate driver. Similarly, Fig. 27 shows the circuit waveforms when the output voltage is regulated to 200V at k=0.37 (or d=45mm) for rated, 50% and 10% load, respectively. For the more tightly coupled system of k=0.37, the resonant inductor current is higher as the short circuit current in the pickup inductor is higher. In addition, the peak capacitor (or switch) voltage is also higher. This is because a higher clamp time is needed for the higher coupling condition to regulate the output at 200V. The measured result matches the results in (Wu et al., 2011b). Consequently, a higher clamp time for the same regulated output voltage results in higher peak voltages across the switches as identified in (Wu et al., 2011b). Notably the peak switch voltage does not exceed 450V which is half of the ratings of the physical devices chosen.

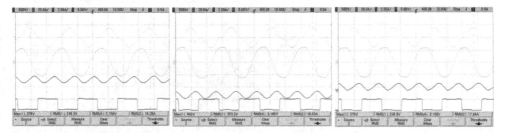

Fig. 26. Measured Waveforms at 200V output with k=0.16 for (a) Rated Load, (b) 50% Load and (c) 10% Load.

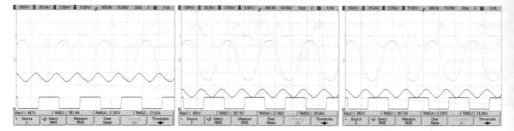

Fig. 27. Measured Waveforms at 200V output with k=0.37 for (a) Rated Load, (b) 50% Load and (c) 10% Load.

The measured output voltage under different operating conditions is shown in Fig. 28. Here, the circuit is tested over a wide range of coupling and load conditions. It can be seen that when the coupling and load resistance are higher than the designed minimum values specified (i.e. $k>0.16$ and $R_L>40\Omega$), the output voltage can be regulated at 200V. However, if a higher load is used (i.e. $R_L<40\Omega$), the output voltage drops below 200V as the minimum clamp time will not be able to produce enough output current to sustain the voltage. Similarly, when $k<0.14$, the output voltage also drops below 200V.

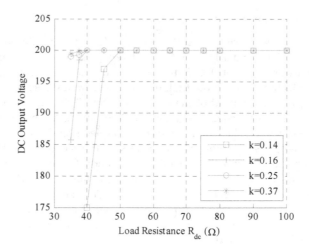

Fig. 28. Output Voltage under Different Circuit Conditions.

The measured efficiency of the secondary AC processing pickup alone is shown in Fig. 29(a). This measurement includes all the losses in the secondary pickup controller and neglects both the losses in the primary converter and the pickup pads. The efficiency of the secondary pickup electronics is as high as 97% when outputting 1kW of controlled output power. As expected the efficiency decreases at lighter loads but as noted there is a point where it drops to a local minimum. This is mainly because the conduction losses in the AC switch are increased as the RMS current in these switches gets higher with increasing clamp time. At much lighter loads, the switch loss decreases because the overall resonant current in the parallel LC network is much lower. The DC-DC efficiency for the IPT system was measured and shown in Fig. 29(b). The efficiency is taken by measuring and comparing the

DC input power to the primary LCL configured power supply used here and the DC output power to the load. As shown, the efficiency drops off at higher coupling conditions because the AC processing pickup reflects higher reactive VAR's back into the primary converter causing more conduction losses in both the LCL network and the H-bridge, but is between 73-83%at rated power, depending on the coupling condition.

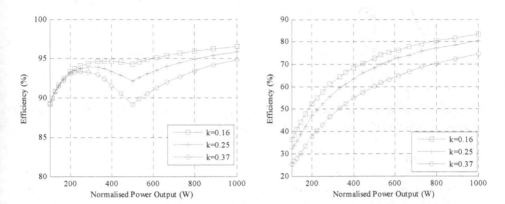

Fig. 29. (a) Efficiency of Parallel AC Processing Pickup, (b) Efficiency of Whole IPT System.

3. Series AC processing pickups

The series AC processing pickup is similar to the parallel AC processing controller described in earlier sections and most of its properties are duals or opposites. As expected it operates as a controllable voltage source and similar to other series IPT pickup designs, it can be very efficient when powering high current low-voltage loads. This section introduces the use of an AC switch operating under Zero-Current-Switching (ZCS) conditions to regulate the output voltage directly for a series AC processing pickup. This topology can produce a high frequency controlled AC voltage source which is ideal for incandescent lamps where the resistance rises as it warms up. When a rectifier is added to the AC output, the circuit can control the output DC voltage. This modified topology eliminates the need for an extra buck converter stage, improves efficiency and reduces cost. In addition, the circuit has the ability to control and eliminate the transient inrush current during start up (Wu et al., 2011c). Preliminary experimental results show that the pickup regulator operates with high efficiency. Fig. 30 shows a traditional series tuned pickup with a buck converter output stage. The pickup stage comprises a series resonant tank, a bridge rectifier and a DC filter capacitor. The buck converter stage includes the switch, the filter inductor and capacitor, and the flywheel diode. The buck converter is used to regulate the output voltage to a desired voltage equal to or less than the rectified open circuit voltage of the pickup. It also protects the circuit if the load is a short circuit. Although there are many advantages to the series compensated pickup with a buck converter, it is not ideal because of the large number of components required in the circuit. In addition, the hard switching operations in the buck converter reduce efficiency and generate more EMI.

The new AC processing pickup proposed in this chapter can also generate a controlled DC voltage, like the series compensated pickup with a buck converter, however the overall

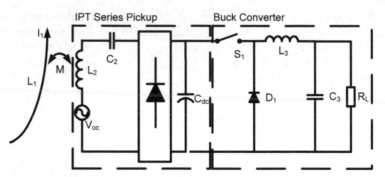

Fig. 30. A series compensated pickup with a buck converter.

circuit requires less components and the efficiency can be higher than that of the traditional circuit as it has soft switching operation, with only one conversion stage.

3.1 Circuit operation

The series AC processing pickup is shown in Fig. 31 with an AC output voltage V_{R2}. Capacitor C_2 is tuned to inductor L_2 at the frequency of the primary track current I_1 to form a series resonant tank. For simplicity, switch S_1 is drawn as an ideal AC switch and is the basis for controlling the output voltage.

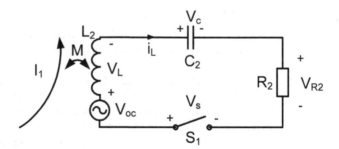

Fig. 31. The Series AC Processing Pickup.

By controlling the switching action of the AC switch, a phase shift ϕ will be generated between the open circuit voltage and the inductor current waveform as shown in Fig. 32. Since the power sourced from V_{oc} is directly proportional to the phase shift ϕ between the voltage and the current, the output power will be directly controlled if the conversion network is assumed to have negligible losses. Hence, by controlling the phase shift ϕ, the output power is controlled.

3.2 Pickup characteristics

The analysis of the series AC processing pickup is very similar to the parallel and can be found in (Wu et al., 2011c). In this section, the operating characteristics of the series AC processing pickup are presented in comparison to the parallel AC processing pickup. The output current (or inductor current) characteristics of the series topology are shown in Fig. 33(a) for different values of Q_{20} or load conditions. The normalized output current is defined

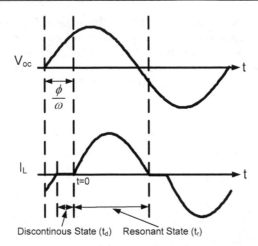

Fig. 32. Waveform showing Two Operating States for Series AC Processing Pickup.

as the ratio of the output current over the short circuit current. For the series AC processing pickup, the output current is dependent on both phase delay ϕ and Q_{20}. The normalized output current can be controlled from a maximum value (or Q_{20}) to zero as ϕ changes for all load conditions. In comparison, for the parallel AC processing pickup, the output current is mostly dependent on ϕ and marginally dependent on Q_{20}.

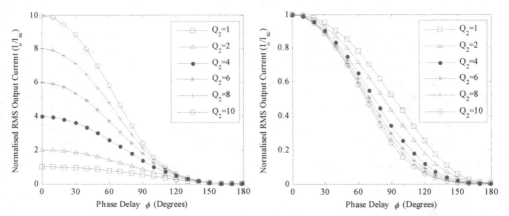

Fig. 33. Normalized RMS Output Current vs. Phase Delay ϕ for (a) Series and (b) Parallel.

The output voltage-current characteristic is shown in Fig. 34(a) for the series topology. The voltage source behaviour is demonstrated as the output voltage stays approximately constant for a given phase delay irrespective of output current as long as the output current is reasonably high. Similarly, the parallel AC processing pickup current-voltage characteristic, shown in Fig. 34(b), appears as a controlled current source. In applications where high currents are required for a short amount of time, the voltage source characteristic is particularly useful. If a parallel topology is used in such applications, it is usually more costly and inefficient to overrate the inductor coil in the parallel AC processing pickup to meet this demand. As previously noted, the series AC processing pickup will give rise to significant advantages in the overall system, as the output voltage can be controlled to any value below the open circuit voltage without the need of an extra buck converter after the pickup (shown in Fig. 30). This eliminates both the losses of an extra DC-DC conversion stage and the need for a costly DC inductor used in the buck converter.

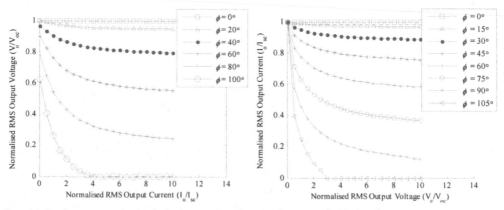

Fig. 34. Pickup Output Characteristics showing (a) Series and (b) Parallel.

3.3 Discussion and experimental results

A series AC processing pickup as described above was constructed with parameters in (Wu et al., 2011c). Fig. 35 shows the circuit waveforms for this series AC processing pickup at 100%, 50% and 20% power when ϕ is set to 0°, 58°, and 85°, respectively. It shows the various operating waveforms: inductor current, capacitor voltage, inductor voltage and switch voltage. The inductor current and capacitor voltage are both sinusoidal having low distortion at 100% power. The small voltage overshoots across the switch result from snubber oscillations as discussed in (Wu et al., 2011c). The measurements as shown have excellent correlation with analytically calculated waveforms in (Wu et al., 2011c) with amplitudes that are within 10% of the calculated values. Since the series AC processing pickup has an AC voltage source property, it can be used to drive incandescent lights.

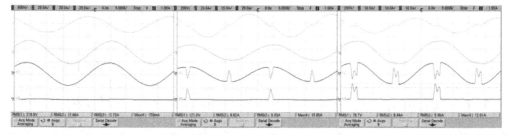

Fig. 35. Measured Waveforms for (a) 100% power, (b) 50% power and (c) 20% power with light bulb load.

An efficiency vs. output power plot is shown in Fig. 36(a) for the series AC processing pickup outputting a controlled AC voltage to a 6Ω load. In addition, the efficiency of the overall system including the AC processing pickup with rectifier outputting controlled DC to an 8Ω load is also plotted. The overall IPT system efficiency is determined using measurements of the DC input power to an LCL primary converter used as the IPT source, through to the AC output power from the secondary pickup. Similarly, the pickup efficiency is calculated using the AC input power delivered by the open circuit voltage source (V_{oc}) and the AC output power from the secondary pickup. The V_{oc} source is only present in a transformer coupling model and cannot be directly measured while the circuit is in operation. As such, an extra coil L_4 is used to estimate the actual phase and magnitude of V_{oc}. This method was found to be an accurate technique for measuring the input power given by:

$$P_{in} = \frac{\omega}{2\pi} \int_0^{2\pi/\omega} V_{oc}(t) I_L(t) dt \qquad (27)$$

The pickup efficiency measurement neglects the supply (LCL converter) power losses and gives a more meaningful measure of the conversion efficiency of the pickup itself. The DC efficiency is measured with the losses in the rectifier taken into account. The efficiency of the pickup remains above 91% when the output power is more than 100W to the AC load. With a 1.2kW load, the efficiency of the series AC processing pickup and the overall IPT system can reach as high as 93% and 84%, respectively when outputting either controlled AC or DC voltage. A breakdown of the power loss measurements for the series AC processing pickup is shown in Fig. 36(b). The switch losses, pickup inductor coil and core losses, capacitor ESR losses, snubber losses, rectifier losses and stray losses are all accounted for individually. The stray loss includes copper losses of the PCB tracks and connectors. At full power there are 35W of loss in the switch giving a 3% reduction in efficiency compared to a traditional series tuned pickup without a buck converter. In a cascaded buck converter, the efficiency would be lower due to additional losses in the DC inductor. At lower power levels, the long discontinuous parts in the resonant waveforms as a result of the controlled switch operation (Fig. 32) decrease the losses in all components and the efficiency of the overall pickup is maintained relatively high. For the harmonic components generated by the discontinuous

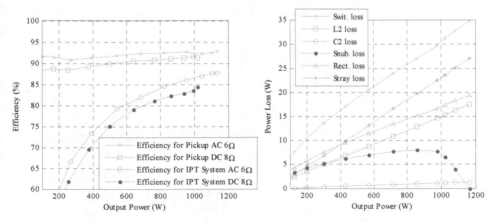

Fig. 36. (a) Efficiency vs. Output Power and (b) Losses vs. Output Power.

control of the AC processing method shown in (Wu et al., 2011c), there are pickup inductor losses due to skin effect. However, the magnitude of the skin effect losses is only 10% of the maximum losses of the fundamental component for this 20kHz IPT system.

4. Conclusions

This chapter has presented two new IPT pickups which have significant advantages compared to traditional pickups that use AC-DC-AC conversion topologies for producing a controllable AC output voltage. Both simple AC processing pickup controllers described can provide controlled AC power over a wide resistive load range. In addition, the simple pickup circuitry has advantages in terms of increasing system efficiency and reducing the pickup size. The fundamental properties and design equations of both the series and parallel tuned AC processing pickup have been investigated.

5. References

Wu, H.H.; Covic, G.A.; Boys, J.T.; Ren, S. & Hu, A.P. (2009a). An AC Processing Pickup for IPT Systems. *Proceedings of the IEEE Energy Conversion Congress and Expo*, ISBN 978-1-4244-2893-9, San Jose, USA, September 2009

Wu, H.H.; Covic, G.A. & Boys, J.T. (2010). An AC Processing Pickup for IPT Systems. *IEEE Transactions on Power Electronics*, Vol. 25, No. 5, (2010), pp. 1275-1284, ISSN 0885-8993

Boys, J.T.; Covic, G.A. & Green, A.W. (2000). Stability and Control of Inductively Coupled Power Transfer Systems. *IEE Proceedings on Electric Power Applications*, Vol. 147, No. 1, (2000), pp. 37-43, ISSN 1350-2352

Elliot, G.A.J. ; Covic, G.A. ; Kacprzak, D. & Boys, J.T. (2006). A New Concept: Asymmetrical Pick-ups for Inductively Coupled Power Transfer Monorail Systems. *IEEE Transactions on Magnetics*, Vol. 42, No. 10, (2006), pp. 3389-3391, ISSN 0018-9464

Wu, H.H.; Covic, G.A.; Boys, J.T & Ren, S. (2009b). Analysis and Design of an AC Processing Pickup for IPT Systems. *Proceedings of the 35th Annual Conference of IEEE Industrial Electronics*, ISBN 978-1-4244-4648-3, Porto, Portugal, November 2009.

Wu, H.H.; Covic, G.A.; Boys, J.T. & Robertson, D. (2011a). A Practical 1.2kW Inductive Power Transfer Lighting System using AC Processing Controllers. *To be published on Proceedings of the IEEE Conference on Industrial Electronics and Applications*, Beijing, China, June 2011

Wang, C.-S.; Stielau O.H. & Covic, G.A. (2005). Design Considerations for a Contactless Electric Vehicle Battery Charger. *IEEE Transactions on Industrial Electronics*, Vol. 52, No. 5, (2005), pp. 1308-1314, ISSN 0278-0046

Budhia, M.; Covic, G.A. & Boys, J.T. (2009). Design and Optimisation of Magnetic Structures for Lumped Inductive Power Transfer systems. *Proceedings of the IEEE Energy Conversion Congress and Expo*, ISBN 978-1-4244-2893-9, San Jose, USA, September 2009

Wu, H.H.; Covic, G.A. & Boys, J.T. (2011b). A 1kW Inductive Charging System using AC Processing Pickups. *To be published on Proceedings of the IEEE Conference on Industrial Electronics and Applications*, Beijing, China, June 2011

Wu, H.H.; Covic, G.A.; Boys, J.T. & Robertson, D. (2011c). A Series Tuned Inductive Power Transfer Pickup with a Controllable AC Voltage Output. *IEEE Transactions on Power Electronics*, Vol. 26, No. 1, (2011), pp. 98-109, ISSN 0885-8993

12

Maximizing Efficiency of Electromagnetic Resonance Wireless Power Transmission Systems with Adaptive Circuits

Huy Hoang and Franklin Bien
Ulsan National Institute of Science and Technology
South Korea

1. Introduction

Wireless power transmission (WPT) is a cutting-edge technology that signifies a new era for electricity without a bunch of wires. Wireless power or wi-power is increasingly becoming the main interest of many R&D firms to eliminate the "last cable" after the wide public exposure of Wi-Fi lately. Even though the first idea was devised from Nikola Tesla in the early 20th century (Tesla, 1919), there was never strong demand for it due to the lack of portable electronic devices. In recent years, with the advent of a booming development in cell-phones and mobile devices, the interest of wireless energy has been re-emerged. WPT offers the possibility to supplying power for electronic devices without having to plug them into AC socket. Until now, there have been many efforts to be made to improve this technology as well as its applications. These efforts include medium-range transmission based on electromagnetic resonance and long-range transmission using microwaves (Greene et al., 2007; Brown & Eves, 1992). Although the investigation of long-range power delivery via far-field techniques was carried out with endeavors, the efficiency or power delivery is still quite low that is not sufficient to fully charge typical electronic gadgets overnight. Therefore, increasing the transmitting power is necessary to provide energy enough to consistent DC supply of gadgets. However, the system would be harmful to human according to IEEE standard for radio frequency electromagnetic fields (IEEE, 1999). The other way is to utilize many transmitters simultaneously, but the implementation seems to be impractical. Additionally, the existence of an uninterruptible line of sight (LoS) is mandatory for microwave-based power transmissions and in a case of mobile objects requiring a complicated tracking system. In general, such power transmissions are relatively suited to very low power applications unless they are used in military or space explorations which are less regulated environments. On the other hand, medium-range WPT covering up to 30 feet is a growing research area that finds wide applications. In order to implement a viable WPT system, Q factors of coils and an impedance matching issue are critical and sensitive to achieve a high efficiency. In reality, due to a resonant coupling nature of the system, for the most efficient power transmission, there is an optimum range between a power coil and a transmitting coil for a fixed distance between the transmitting and receiving coils. This effect may not be clarified by a conventional magnetic induction theory. In this chapter, a simple equivalent circuit model for a WPT system via electromagnetic resonance will be derived and analytically solved. From the solution, above effect could be easily clarified and key concepts including frequency splitting and impedance matching will be mentioned as well. In addition, adaptive circuits, called antenna-locked loops (ALL), are studied to maintain the optimal resonant condition, realize the maximum wireless power transfer efficiency and execute precision resonant frequency optimization by setting the resonant condition with the low Q factor to detect any possible incoming power, then increasing the Q factor while maintaining the resonant condition.

This chapter is organized as follows. In Section 2, Subsection 2.1 introduces a system model and circuit analysis of a four-coil system. Subsection 2.2 describes a comparison of different types of coupling mechanism, while a case of multiple receivers in WPT system is mentioned in Subsection 2.3. Subsection 2.4 shows experimental results. The ideas for ALL systems are presented in Section 3. Finally, Section 4 provides conclusion.

2. System model and circuit analysis

The electromagnetic resonance (also called magnetic resonant coupling) WPT based techniques are typically relied on four coils as opposed to two coils used in the conventional inductive links. A typical model of four-coil power transfer system is shown in Fig. 1, which consists of a power coil, a transmitting coil (Tx coil), a receiving coil (Rx coil) and a load coil. The transmitting coil and the receiving coil are also called resonators, which are supposed to resonate at the same frequency. For common cases, sizes of the four coils are different. Indeed, in some applications, the coils in the receiver side are needed to be scaled as small enough to be integrated in portable devices such as laptops, handheld devices or implantable medical equipment. In various cases of practical interest, the receiving and load coils can be fitted within the dimensions of those personal assistant tools, enabling mobility and flexibility properties. Otherwise, it is quite free to determine sizes of the transmitter. Normally, the transmitting coil may be made larger for a higher efficiency of the system. For the system in Fig. 1, a drawback of a low coupling coefficient between the Tx and Rx coils, as they locate a distance away from each other, is possibly overcome by using high-Q coils. This may help improve the system performance. In other words, the system is able to maintain the high efficiency even when the receiver moves far away from the transmitter. In the transmitting part, a signal generator is used to generate a sinusoidal signal oscillating at the frequency of interest. A power of the output signal from the generator is too small,

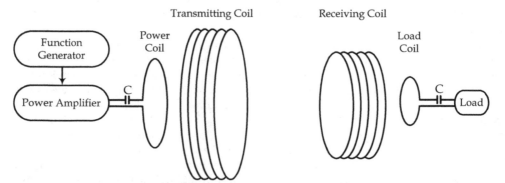

Fig. 1. Model of wireless power transfer system.

approximately tens to hundreds of milliwatts, to power devices of tens of watts. Hence, this signal is delivered to the Tx coil through a power amplifier for signal power amplification. In the receiver side, the receiving resonator and then load coil will transfer the induced energy to a connected load such as a certain electronic device. While the efficiency of the two-coil counterpart is unproportionally dependent on an operating distance, the four-coil system is less sensitive to changes in the distance between the Tx and Rx coils. This kind of system can be optimized to provide a maximum efficiency at the given operating distance. These characteristics will be analyzed in the succeeding sections.

2.1 Circuit analysis of four-coil system

Fig. 2 shows the circuit representation of the four-coil system as modeled above. The schematic is composed of four resonant circuits corresponding to the four coils. These coils are connected together via a magnetic field, characterized by coupling coefficients k_{12}, k_{23}, and k_{34}. Because the strengths of cross couplings between the power & Rx coils and the load & Tx coils are very weak, they can be neglected in the following analysis. Theoretically, the coupling coefficient (also called coupling factor) has a range from 0 to 1. If all magnetic flux generated from a transmitting coil is able to reach a receiving coil, the coupling coefficient would be "1". On the contrary, the coefficient would be represented as "0" when there is no interaction between them. Actually, there are some factors identifying the coupling coefficient. It is effectively determined by the distance between the coils and their relative sizes. It is additionally determined by shapes of the coils and orientation (angle) between them. The coupling coefficient can be calculated by using a given formula

$$k_{xy} = \frac{M_{xy}}{\sqrt{L_x L_y}} \tag{1}$$

where M_{xy} is mutual inductance between coil "x" and coil "y" and note that $0 \le k_{xy} \le 1$. Referring to the circuit schematic, an AC power source with output impedance of R_s provides energy for the system via the power coil. Normally, the AC power supply can be either a power amplifier or a vector network analyzer (VNA) which is useful to measure a transmission and reflection ratio of the system. Hence, a typical value of R_S, known as the output impedance of the power amplifier or the VNA, is $50\,\Omega$. The power coil can be modeled as an inductor L_1 with a parasitic resistor R_1. A capacitor C_1 is added to make the power coil resonate at the desirable frequency. The Tx coil is a helical coil with many turns represented as an inductor L_2 with parasitic resistance R_2. Geometry of the Tx coil

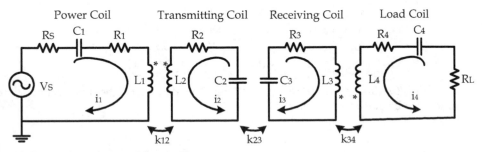

Fig. 2. Equivalent circuit of four-coil system.

determines its parasitic capacitance such as stray capacitance, which is represented as C_2. Since this kind of capacitance is difficult to be accurately predicted, for fixed size of the coil, a physical length, which impacts the self-inductance and the parasitic capacitance, has been manually adjusted in order to fit the resonant frequency as desired. In the receiver side, the Rx coil is modeled respectively by L_3, R_3 and C_3. The load coil and the connected load are also performed by L_4, R_4 and R_L. A capacitor C_4 also has the same role as C_1, so that the resonant frequency of the load coil is defined. When the frequency of sinusoidal voltage source V_S is equal to the self-resonant frequency of the resonators, their impedances are at least. In the other words, currents of the coils would be at the most and energy can be delivered mostly to the receiving coil. Otherwise, energy of the transmitting power source would be dissipated in the power coil circuit itself, resulting in the very low efficiency. In general, setting the frequency of AC supply source as same as the natural resonant frequency of the transceiver coils is one of key points to achieve a higher performance of the system.

As can be seen from Fig. 2, the Tx coil is magnetically coupled to the power coil by the coupling coefficient k_{12}. In fact, the power coil is one of the forms of impedance matching mechanism. The same situation experiences in the receiving part where the Rx coil and load coil are magnetically linked by k_{34}. The strength of interaction between the transmitting and receiving coils is characterized by the coupling coefficient k_{23}, which is decided by the distance between these coils, a relative orientation and alignment of them. In general, it is able to use other mechanisms for the impedance matching purpose in either or both sides of the system. For example, a transformer or an impedance matching network, which consists of a set of inductors and capacitors configured to connect the power source and the load to the resonators, is routinely employed. Similar to aspects mentioned above, in reality, the power and Tx coils would be implemented monolithically for the sake of convenience; hence the coupling coefficient k_{12} would be stable. For the same objective, k_{34} would also be fixed. Therefore, there only remains coefficient k_{23} which is so-called an environment variable parameter. The parameter varying with usage conditions, may include the range between the resonator coils, a relative orientation and alignment between them and a variable load on the receiving resonator.

The circuit model offers a convenient way to systematically analyze the characteristic of the system. By applying circuit theory Kirchhoff's Voltage Law (KVL) to this system, with the currents in each resonant circuit chosen as illustrated in Fig. 2, a relationship between currents through each coil and the voltage applied to the power coil can be captured as a following matrix

$$\begin{bmatrix} V_S \\ 0 \\ 0 \\ 0 \end{bmatrix} = \begin{bmatrix} Z_1 & j\omega M_{12} & 0 & 0 \\ j\omega M_{12} & Z_2 & -j\omega M_{23} & 0 \\ 0 & -j\omega M_{23} & Z_3 & j\omega M_{34} \\ 0 & 0 & j\omega M_{34} & Z_4 \end{bmatrix} \begin{bmatrix} i_1 \\ i_2 \\ i_3 \\ i_4 \end{bmatrix} \qquad (2)$$

where Z_1, Z_2, Z_3, and Z_4 respectively are loop impedances of the four coils. These impedances can be indicated as below

$$Z_1 = R_S + R_1 + j\left(\omega L_1 - \frac{1}{\omega C_1} \right) \qquad (3)$$

$$Z_2 = R_2 + j\left(\omega L_2 - \frac{1}{\omega C_2}\right) \tag{4}$$

$$Z_3 = R_3 + j\left(\omega L_3 - \frac{1}{\omega C_3}\right) \tag{5}$$

$$Z_4 = R_L + R_4 + j\left(\omega L_4 - \frac{1}{\omega C_4}\right) \tag{6}$$

From the matrix (2), by using the substitution method, the current in the load coil resonant circuit is derived as given

$$i_4 = -\frac{j\omega^3 M_{12}M_{23}M_{34}V_S}{Z_1Z_2Z_3Z_4 + \omega^2 M_{12}^2 Z_3 Z_4 + \omega^2 M_{23}^2 Z_1 Z_4 + \omega^2 M_{34}^2 Z_1 Z_2 + \omega^4 M_{12}^2 M_{34}^2} \tag{7}$$

It is clearly seen that the voltage across the load is equal to $V_L = -i_4 R_L$ and the relationship between the voltages of source and load is given as V_L/V_S.

The system model can be considered as a two port network. To analyze a figure of merit of this kind of system, S – parameter is a suitable candidate. Actually, S_{21} is a vector referring to a ratio of signal exiting at an output port to a signal incident at an input port. This parameter is really important because a power gain, the critical factor determining of power transfer efficiency, is given by $|S_{21}|^2$, the squared magnitude of S_{21}. The parameter of S_{21} is calculated by (Sample et al., 2011, as cited in Fletcher & Rossing, 1998; Mongia, 2007)

$$S_{21} = 2\frac{V_L}{V_S}\left(\frac{R_S}{R_L}\right)^{1/2} \tag{8}$$

Thus, combining with $M_{xy} = k_{xy}\sqrt{L_x L_y}$ derived from (1), the S_{21} parameter is given as

$$S_{21} = \frac{j2\omega^3 k_{12}k_{23}k_{34}L_2 L_3\sqrt{L_1 L_4 R_S R_L}}{Z_1 Z_2 Z_3 Z_4 + k_{12}^2 L_1 L_2 Z_3 Z_4 \omega^2 + k_{23}^2 L_2 L_3 Z_1 Z_4 \omega^2 + k_{34}^2 L_3 L_4 Z_1 Z_2 \omega^2 + k_{12}^2 k_{34}^2 L_1 L_2 L_3 L_4 \omega^4} \tag{9}$$

It is helpful to analyze the performance of the system according to equation (9). With all the circuit parameters provided in Table 1, the parameter regarded as the factor determining the efficiency of the system, magnitude of S_{21}, can be performed by a function of only two variables k_{23} and frequency. As referred, the coupling coefficient k_{23} is the parameter which varies according to changes in circumstances. A changeable distance, for instance, is a cause of k_{23} variation. In addition, changes in the orientation or misalignment between the transmitting and receiving resonators make the above coefficient inconsistent as well. Actually, when the distance increases, k_{23} will go down because the mutual inductance between those coils declines with distance. In case of a variable orientation or misalignment, the k_{23} also changes. The relation among $|S_{21}|$, k_{23} and frequency is demonstrated in Fig. 3. Note that in practice, a vector of S_{21} parameter including magnitude and phase information can be measured by using VNA. From Fig. 3, it is clearly seen that when k_{23} is small in cases of the large distance between the transmitter and the receiver or the misalignment, orientation deviation taking place, the efficiency represented as S_{21} magnitude is able to

Transmitter Side		Receiver Side	
Parameter	Value	Parameter	Value
R_S	50 Ω	L_3	0.4 μH
L_1	0.5 μH	R_3	0.02 Ω
R_1	0.015 Ω	C_3	357.5 pF
C_1	286 pF	k_{34}	0.1
k_{12}	0.05	L_4	0.1 μH
L_2	1.3 μH	R_4	0.012 Ω
R_2	0.03 Ω	C_4	1.43 nF
C_2	110 pF	R_L	50 Ω
k_{23}	0.0001 to 0.3	frequency	11-16 MHz

Table 1. An example of circuit values.

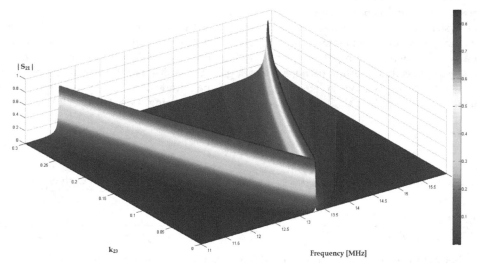

Fig. 3. $|S_{21}|$ as a function of k_{23} and frequency (3D – View).

reach a peak at the self resonant frequency of approximately 13.3 MHz. However, the resonant frequency separates as k_{23} is over a certain level. The phenomenon is so-called frequency splitting which has a negative impact on the system efficiency. For instance, as long as the transmitting and receiving coils are such closed as the coupling coefficient k_{23} between them is 0.1, the resonant frequency splits into two peaks at 12.69 and 14.03 MHz as observed from Fig. 3. Consequently, the system performance is considerably degraded. In order to overcome the drawback, an automatically frequency tuning circuit is proposed, as presented in Section 3. The circuit is used to track the resonant frequency of interest so as to preserve the efficiency of the system in cases of transceivers' mobility. It is possible to simulate the system by using Advanced Design System (ADS) of Agilent Technologies. With the circuit setup illustrated in Fig. 4, the result of the magnitude of S_{21} can be obtained as shown in Fig. 5.

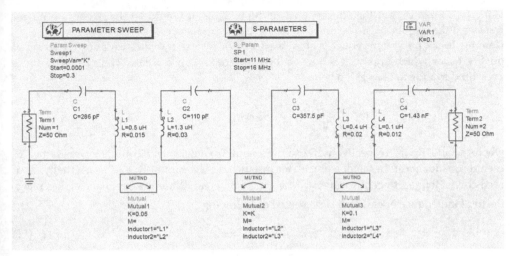

Fig. 4. Simulation setup using Advanced Design System (ADS).

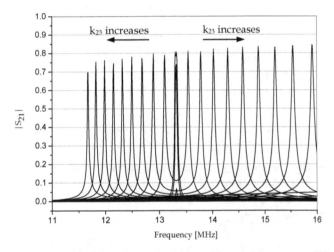

Fig. 5. Simulation result showing $|S_{21}|$ as a function of k_{23} and frequency (2D – View).

It is instructive to analyze carefully a trend of $|S_{21}|$ as k_{23} variation. Fig. 5 clarifies that when the coefficient k_{23} is absolutely small corresponding to a case that the transmitter and the receiver are too far away each other, $|S_{21}|$ is low. When the distance between the resonators is getting closer, k_{23} increases bringing about a higher magnitude of S_{21}. However, as $|S_{21}|$ increases to a certain level, the higher k_{23} does not lead to the higher amount of $|S_{21}|$. Moreover, there is the frequency splitting issue which substantially reduces the system efficiency. The point, at which the deviation of the original resonant frequency (13.3 MHz) happens, plays a prominent role in the system. It clarifies the relative position of the resonators that the performance of the system is the highest. If the distance is longer than that range, the efficiency is poorly defined. On the contrary, the resonant frequency detunes along two furrows, but the efficiency is still high. Thus, it would be the maximum power transfer if the frequency can be tuned to the desirable frequency.

Coming back the system equation indicated in (9), let expand this equation in terms of quality factor which appreciates how well the resonator can oscillate. The quality factor is presented in a formula as given below

$$Q_i = \frac{1}{R_i}\sqrt{\frac{L_i}{C_i}} = \frac{\omega_i L_i}{R_i} \Leftrightarrow \omega_i L_i = R_i Q_i, \, i = 1 \sim 4 \tag{10}$$

where ω_i and R_i are respectively the self-resonant frequency and equivalent resistance of each resonant circuit. In the power coil, for instance, R_i is a sum of R_S and R_1. Actually, ω_i of each coil is defined to be the same, $\omega_1 = \omega_2 = \omega_3 = \omega_4 = \omega_0$. When the resonance takes place, the total impedance of each coil is presented as following

$$Z_1 = R_S + R_1 \approx R_S \tag{11}$$

$$Z_2 = R_2 \tag{12}$$

$$Z_3 = R_3 \tag{13}$$

$$Z_1 = R_L + R_4 \approx R_L \tag{14}$$

For simplicity, in addition to the fact that system parameters can be measured by VNA, it is common to set R_S equal to R_L. At the resonant frequency, $\omega_0 = 1 / \sqrt{L_i C_i}$, from (9), the magnitude of S_{21} can be written as

$$|S_{21}| = \frac{2k_{12}k_{23}k_{34}Q_2Q_3\sqrt{Q_1Q_4}}{1 + k_{12}^2Q_1Q_2 + k_{23}^2Q_2Q_3 + k_{34}^2Q_3Q_4 + k_{12}^2k_{34}^2Q_1Q_2Q_3Q_4} \tag{15}$$

As referred previously, the coupling coefficient k_{12} and k_{34} would be constant. There is only k_{23} varying with medium conditions. To find the range between the resonators at which $|S_{21}|$ or the efficiency is certainly at maximum, a derivative of S_{21} with respect to k_{23} is taken and then setting the result to zero, yielding

$$\frac{d|S_{21}|}{dk_{23}} = 0 \Rightarrow k_{23}^* = \sqrt{\frac{\left(1 + k_{12}^2Q_1Q_2\right)\left(1 + k_{34}^2Q_3Q_4\right)}{Q_2Q_3}} \tag{16}$$

This value of k_{23}^* is equivalent to the maximum range that the transmitter is able to effectively transfer power to the receiver at the given resonant frequency (before the resonant frequency breaking in two peaks). Note that $k_{23}^* \leq 1$. With the purpose of finding out the maximum efficiency of the system in terms of $|S_{21}|$, it is feasible to substitute k_{23}, which is derived above, into equation (15)

$$|S_{21}|_{max} = \frac{k_{12}k_{34}Q_1Q_4R_L}{k_{23}^*\sqrt{L_1\omega_1L_4\omega_4}} = \frac{k_{12}k_{34}Q_1Q_4R_L}{k_{23}^*\sqrt{L_1L_4}\omega_0} \tag{17}$$

It is clear that $|S_{21}|_{max}$ unproportionally depends on k_{23}^{*}. It means for the sake of a higher efficiency, the extent that the highest efficiency can be achievable is shortened. In order to get a greater value of $|S_{21}|_{max}$, k_{23}^{*} is supposed to decrease. From equation (16), increasing Q_2 and Q_3 is able to reduce the k_{23}^{*}. In general, making the very high-Q transmitting and receiving coils is very crucial so as to achieve the high transfer performance.

For example, from equation (17), with the value given in Table 1, the maximum value of magnitude of S_{21} parameter is calculated as follows

$$\omega_0 = \omega_1 = \frac{1}{\sqrt{L_1 C_1}} \approx 83.624 \times 10^6 \ [rad / s]$$

$$Q_1 = \frac{\omega_0 L_1}{R_S + R_1} \approx 0.84$$

$$Q_2 = \frac{\omega_0 L_2}{R_2} \approx 3623.71$$

$$Q_3 = \frac{\omega_0 L_3}{R_3} \approx 1672.48$$

$$Q_4 = \frac{\omega_0 L_4}{R_L + R_4} \approx 0.17$$

$$k_{23}^{*} = \sqrt{\frac{\left(1 + k_{12}^2 Q_1 Q_2\right)\left(1 + k_{34}^2 Q_3 Q_4\right)}{Q_2 Q_3}} \approx 2.34 \times 10^{-3}$$

$$|S_{21}|_{max} = \frac{k_{12} k_{34} Q_1 Q_4 R_L}{k_{23}^{*} \sqrt{L_1 L_4} \omega_0} \approx 0.82$$

2.2 Different coupling mechanism systems in wireless power transfer

As mentioned in Subsection 2.1, the advantage of the four coils system over the two coils system is a high efficiency even in far afield condition. Why is that so? To answer this question, it is instructive to study three different coupling mechanism based circuits which are demonstrated in Fig. 6. A non-resonant inductive coupling circuit in Fig. 6(a) is totally based on the principle of an ordinary transformer. This kind of power transfer also uses primary and secondary coils as similar as transformer, but a striking feature is an exclusion of a high permeability coil. Since an energy transmission is relied on the induction principle, more power is dissipated along the coil or ambient environment and it is more difficult to achieve a long distance transmission.

The above limitation can be overcome using the WPT based on resonant coupling shown in Fig. 6(b). By adding external capacitors, coils in primary and secondary side are able to resonate at the same frequency of interest. In fact, high quality factor coils are considered as one of the most critical features for a superior system.

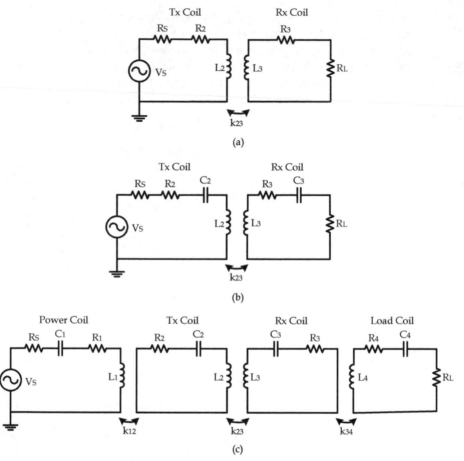

Fig. 6. Three different coupling mechanism circuits.
a.   Non-resonant inductive coupling based circuit.
b.   Low-Q resonant coupling based circuit (two-coil system).
c.   High-Q resonant coupling based circuit (four-coil system).

In case of Fig. 6(b), quality factors of the two resonant circuits are determined by the loading provided by R_S and R_L which are also two major contributors to loss of circuits (Cannon et al., 2009). Source and load resistances are leading causes of lower Q resonators, deteriorating the system efficiency. A solution for this matter is to separate the R_S and R_L from the resonators, that is illustrated in Fig. 6(c). Certainly, the resonators have larger quality factors due to the elimination of the unexpected resistances. It is apparent that the quality factors of the transmitting and receiving coils dominantly affect the system performance. In order to comprehend more deeply about the three different circuits, an example with circuit parameters shown in Table 2 is put forward. Fig. 7 illustrates a comparison result of the three different coupling methods including inductive coupling, low-Q resonant coupling and high-Q resonant coupling. Note that the two resonant coupling circuits resonate at 8 MHz.

Transmitter Side		Receiver Side	
Parameter	Value	Parameter	Value
R_S	50 Ω	L_3	5 μH
L_1	2 μH	R_3	0.7 Ω
R_1	0.4 Ω	C_3	79.2 pF
C_1	198 pF	k_{34}	0.1
k_{12}	0.1	L_4	1 μH
L_2	30 μH	R_4	0.25 Ω
R_2	2 Ω	C_4	396 pF
C_2	13.2 pF	R_L	50 Ω
k_{23}	0.001	frequency	4 – 12 MHz

Table 2. Example of component values for three circuit models.

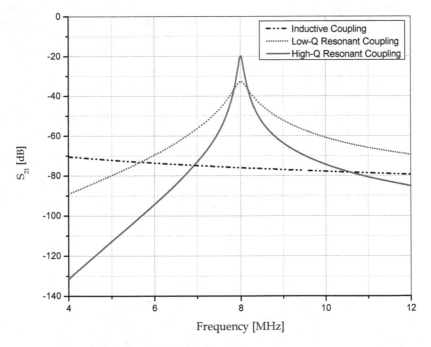

Fig. 7. Comparison result of three different types of coupling.

As can be seen, the value of S_{21} in dB is used for the comparison. It is evident that for the inductive coupling mechanism shown in Fig. 6(a), the parameter of S_{21} is the lowest. In fact, this value gradually declines from -70 dB to about -80 dB for a frequency range between 4 and 12 MHz. By above analysis, the Q factor of the circuit shown in Fig. 6(c) is much greater than that of Fig. 6(b). In fact, from Fig. 7, S_{21} parameter of the high-Q circuit is approximately 20 dB higher than that of the low-Q circuit. That completely proves the theoretical presumption.

2.3 Wireless energy transmission to multiple devices through resonant coupling

All the approaches mentioned previously are merely in terms of one to one WPT. That means one transmitter, which includes a power coil and a transmitting coil, provides energy wirelessly to only one receiver consisting of receiving and load coils in a distance away. In reality, however, the cases of multiple small receivers are in favor and needed to be considered carefully. Transferring power to a couple of receivers is also based on the same principle as one to one case. Nevertheless, an effect of two receivers in proximity is considerable. Thus, several cases of multiple receivers wireless energy transmission will be investigated. In case of two identical receivers located sufficiently far field and there is no interaction between them, the system can be interpreted as a sum of two discrete systems. Since the two receivers are identical, their operations are coincident with each other if they experience a same condition such as the strength of coupling. With the circuit parameters shown in Table 2, only difference in the coupling coefficient between the transmitting and receiving coils, the performance of the two receivers is illustrated in Fig. 8. It is undoubtedly true that the resonant frequency splits into two peaks as an increase of k, which is the coupling coefficient between the two receivers and the transmitter. The stronger the coupling is, the more the new resonant frequencies deviate from the original resonant frequency. At k of 0.01, for example, the system efficiency hits the peak at 8 MHz. When the coupling getting stronger to 0.145 and then 0.3, the original peak respectively breaks in two other peaks at about 7.2 and 9.1 MHz; 6.7 and 10.6 MHz. On the other hand, as shown in Fig. 9, in case of the strong interaction between receiving coils, even at low k, the resonant frequency is splitted to two peaks at 7 and 8 MHz. When k reaches 0.145, the maximum power transfer occurs at the frequency of 6.7 and 8.5 MHz. The separation among splitted frequencies is larger at the stronger coupling between the transmitter and the receivers, 6.3 and 9.6 MHz. For a situation that the two receivers resonate at the same frequency but their

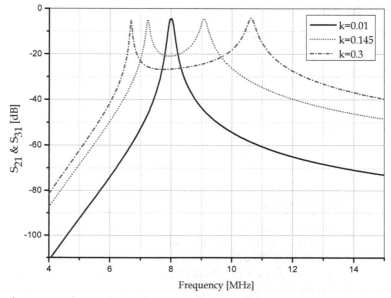

Fig. 8. Performance of two identical receivers in case of no interaction between them.

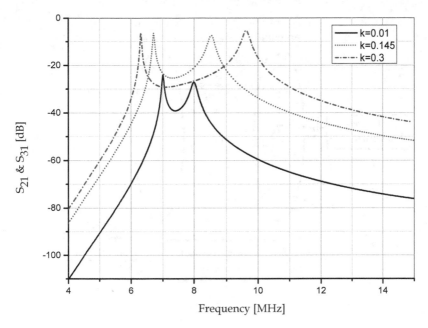

Fig. 9. Performance of two identical receivers in case of strong interaction.

physical parameters are different, the system transfer efficiency is relatively similar. Theoretically, the four circuit model equations derived from the matrix equation (2) can be extended for multiple receivers. For one to two system, in particular, the extension of circuit equations is straightforward, with six equations instead of four. By using these equations, it is possible to predict the characteristic of the system with multiple receivers.

2.4 Experimental results
An experiment, which is conducted for WPT, was presented (Imura & Hori, 2011). The experimental setup is illustrated in Fig. 10. For S – parameter measurement, a transmitter and a receiver of the power transmission system are in turn connected to port 1 and port 2 of VNA. As same as theoretical analysis, the transmitting and receiving antennas resonate at same frequency. The helical antennas used are short– type antennas, which have separate excitation using self-inductances and added capacitors. These antennas have only one turn each with a radius of 150 mm and attached capacitors in series to adjust the resonant frequency of interest. The experiment is conducted with the distance between two antennas respectively 49, 80, 170 and 357 mm. From Fig. 11, at the closed distance of 49 mm, the system achieves the highest efficiency, represented as the squared magnitude of S_{21} parameter, at the two peaks of roughly 12.4 and 15.2 MHz. When the distance is getting smaller, the resonant frequency separation reduces, about 12.7 and 14.6 respectively. And at the distance of 170 mm, the two splitted frequencies converge at approximately 13.6 MHz. The efficiency significantly degrades with the increasing distance. The special point distinct from the presented model of WPT is that there are no power coil and load coil in this model. The authors used two resonators in addition to impedance matching structures instead of

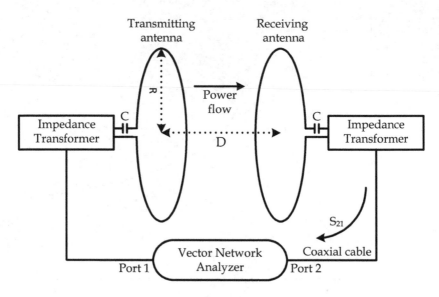

Fig. 10. Experiment setup.

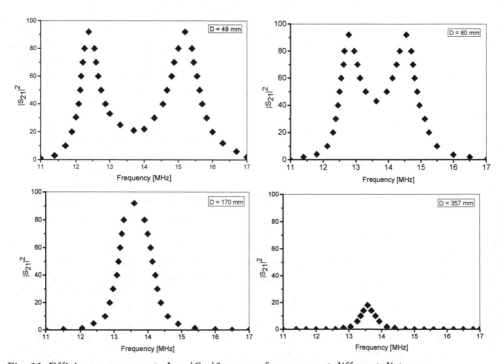

Fig. 11. Efficiency, represented as $|S_{21}|^2$, versus frequency at different distances.

the four-coil system. The advantage of this model is a possibility of eliminating the magnetically coupled coils so that the system would be simplified. Nonetheless, the benefits of cross-coupling effect increasing the system efficiency in the low-mode of resonant frequencies can be cancelled out (Sample et al., 2011).

3. Maximizing efficiency with adaptive circuits

From the above analysis of the relationship between the system efficiency and the resonant frequency, it is clear that the operating frequency is the critical factor determining the performance of the system. Besides, the flexibility of impedance matching structures also plays an important role enabling high transfer efficiency (Chen et al., 2010). Of considerable interest for applications of WPT relied on electromagnetic resonance, the cases of mobile receiver or multiple receivers are absolutely typical. However, there exists a drawback that degenerates the efficiency in these cases. In fact, the transfer efficiency significantly decreases with distance variations between the transmitter and the receiver or in case of multiple receivers. In order to overcome the limitations, adaptive circuits are proposed. These circuits are so-called ALL which help to maintain the optimal resonant condition and realize the maximum wireless power transfer efficiency as well.

3.1 Efficiency optimization based on frequency control

For the situation of one transmitter and one portable receiver, the transfer efficiency represented as $|S_{21}|$, which the function of the distance, the relative orientation and alignment between the resonators, is analytically clarified in the previous section. Remind that magnitude of S_{21} parameter is relatively small when a transmitter and a receiver are too far away. When they get approach each other, $|S_{21}|$ goes up and at a certain point, the phenomenon of frequency splitting occurs degrading the system performance. Therefore, an optimal control mechanism of efficiency based on frequency control is needed to stabilize the transfer efficiency.

Generally, a range of control frequency is confined, with a high limit caused by the coil characteristic and a low limit due to the low efficiency. In that range, the frequency can be determined and tuned in order for high efficiency to be achieved. From the equations (7) and (8), it is possible to derive a following equation

$$S_{21} = \frac{j2\omega^3 M_{12} M_{23} M_{34} R_L}{Z_1 Z_2 Z_3 Z_4 + \omega^2 M_{12}^2 Z_3 Z_4 + \omega^2 M_{23}^2 Z_1 Z_4 + \omega^2 M_{34}^2 Z_1 Z_2 + \omega^4 M_{12}^2 M_{34}^2} \quad (18)$$

In which mutual inductance M_{23} is calculated by using Neumann formula (Imura & Hori, 2011, as cited in Sallan et al., 2009)

$$M_{23} = \frac{\mu_0}{4\pi} \int\limits_{C_2} \int\limits_{C_3} \frac{dl_2 dl_3}{D} \quad (19)$$

However, due to complicated calculations, it is reasonable to use an approximation of the mutual inductance given as below (Karalis et al., 2008 as cited in Jackson, 1999)

$$M_{23} \approx \pi\mu_0 (r_2 r_3)^2 \frac{N_2 N_3}{2D^3} \quad (20)$$

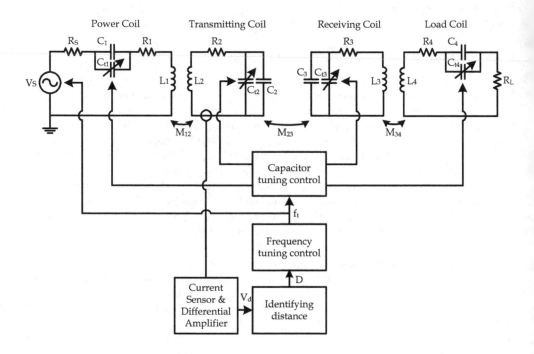

Fig. 12. Adaptive circuit of frequency control.

Note that in (18), typically, almost all components would be identified with given specifications of circuit setup including radius of coils' cross-section a, number of turns N, radius of coils r_i (i=2,3) , and distance between power coil, load coil and resonators. So, by substituting (19) into (18), there are merely the three unknown variables of frequency ω, S_{21} parameter and distance between the resonators D. With the given requirement of efficiency, represented as the magnitude of S_{21}, and identified distance between the resonators, it is able to figure out the frequency of interest. An adaptive circuit used to stabilize the system transfer efficiency is demonstrated in Fig. 12. A current sensor is used to detect a current flow in the transmitting coil. Due to the fact that the transmitting coil is not connected to the ground, the sensed signal is in terms of differential signal. The signal is then compared with reference sources in an adjacent block, hence it is essential to utilize a differential amplifier in order to transform the differential signal to a single-ended signal. An output voltage of V_d is then switched to a block of distance identification, where V_d is in turn compared with reference voltages to determine a distance between the resonators. Like the preceding analysis, with the found parameter, a new tuned f_t is established. This frequency is the wanted frequency of the power source as well. Subsequently, in order to control all coils resonating at the frequency of f_t, a capacitor tuning control block is required to control variable capacitors attached at each coil as below

$$f_t = \frac{1}{2\pi L_i C_{total-i}}, i = 1 \sim 4 \qquad (21)$$

$$C_{ti} = C_{total-i} - C_i, i = 1 \sim 4 \qquad (22)$$

Note that C_2 and C_3 here are lumped components representing approximately the parasitic capacitances of the transmitting and receiving coils. The capacitors C_{ti} with i from 1 to 4 are respectively connected in parallel with the capacitors of four coils.

In general, when the frequency tuning mechanism is enabled, the controller picks the resonant frequency of interest and tracks it as the receiver is moved away from the transmitter.

3.2 Efficiency optimization based on impedance matching control

In addition to the efficiency optimization technique based on frequency tuning, impedance matching tuning method is a potential candidate for an adaptive circuit that also maximizes the system efficiency. In some cases, the usage of wide range of frequency tuning has limitations that can affect these other bands such as ISM bands which were internationally reserved. Thus, utilizing the technique of flexible impedance matching is really essential.

In fact, by changing the strength of coupling between the load coil and the resonator and slightly retuning the receiving coil, it is possible to achieve the maximum transfer efficiency (Chen et al., 2010, as cited in Kurs et al., 2007). For practical interest, however, an adjustment in the coupling coefficient between the coils in the transmitting part is preferred. The change in coupling strength can be made by varying the distance between those coils, the relative orientation and alignment of them. However, it is not viable to automatically control them in the system consisting of four coils. Thus, a model of two resonators and other impedance matching structure is used. A circuit of adaptive impedance matching in the transmitter side is shown in Fig. 13. Based on a current sensed from the transmitting resonator, a control circuit block is able to identify distance variations, different orientation, misalignment between the resonators or in case of multiple receivers, then automatically control a power amplifier (PA) and a tunable impedance matching block so as to maximize the transfer efficiency. Actually, for situations of relatively large distance length, significantly different orientation or misalignment between the two resonators, in spite of utilizing the adaptive impedance matching, increasing the output power of the power amplifier is recommended to improve the system transfer efficiency. The striking feature of the circuit is that the system frequency is fixed and it is very helpful in many applications.

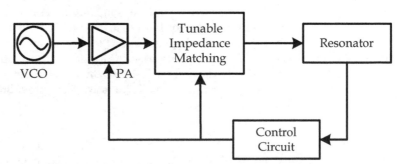

Fig. 13. Adaptive circuit of impedance matching control in transmitter side

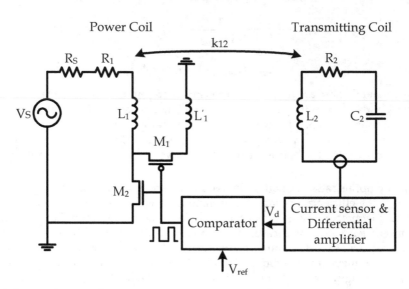

Fig. 14. Adaptive circuit of Q-tuning.

3.3 Adaptive Q tuning circuit

From the analysis of different coupling mechanisms in Subsection 2.2, the high-Q magnetic resonant coupling provides the best transfer performance of system. In some applications, however, the low-Q magnetic coupling system has its own advantages. As seen from Fig. 7 which shows the comparison between three kinds of coupling, despite the lower efficiency, the low-Q coupling mechanism operates in a wide range of frequency rather than the high-Q coupling. That is why the low-Q factor can be used to detect any possible incoming power. An adaptive Q-tuning circuit is illustrated in Fig. 14. There is no added capacitor in the power coil to set the expected frequency. Actually, the power coil inductively couples with the transmitting coil. Regarding the power coil, L_1 has a number of turns N_1 and L_1' with N_1'. These coil turns are connected together by a switch which is implemented by a power MOSFET M_1. As same as the previous explanation in Subsection 3.1, a current sensor and a differential amplifier are used. In case of the detection of either an absence of any receivers or multiple of them, the sensed current is low causing a small value of V_d. This value is then compared with a reference voltage V_{ref} by a comparator. Because of the lower value of V_d than V_{ref}, the output of the comparator is set to low, which makes the switch M_1 turn on while M_2 turn off. By the way, the number of coils turns increase by N_1', which reduces the quality factor Q of the transmitting resonator due to a lower turns ratio between the power coil and the transmitting coil (Cannon et al., 2009). On the other hand, in case of one to one system, the switch of M_1 is degenerated while M_2 is activated providing a higher turns ratio which raises the Q factor of the resonator.

4. Conclusion

A general and insightful analysis of WPT system based on electromagnetic resonance is presented. Frequency splitting phenomenon is demonstrated by theoretical derivations and simulation results as well. Besides, the comparison between different kinds of coupling and case of multiple receivers are also analyzed to impress the need for adaptive circuits to maintain the high performance of the system. Called Antenna- Locked Loops, these circuits offer practical possibilities of WPT with any physical changes. With the wireless power know-how, it is able to counter the transmission of power over distances about tens of feet, although ideally it is very less but still it is impressive. The most interesting fact is that the wireless power transmission is omni directional in nature. If the technology is enhanced and sharpened to be a datum where it can be "generative", it will be able to remain firm to turn the interest of an infinite number of industries. Although, nowadays wireless power is a major obstacle in terms of advancement in the retail sector and also there are many issues regarding the safety, applying and affordability in attentiveness to WPT, but this will likely to be enhanced as the technology further grows up. Generally, this work lays down the ground work of innovative wireless power technology and open opportunities to commercially implement advanced electromagnetic resonance based WPT systems.

5. Acknowledgment

This work was supported by Basic Science Research Program through the National Research Foundation of Korea (NRF) funded by the Ministry of Education, Science and Technology (grant number 20110005518).

6. References

Brown, W. C. & Eves, E. E. (1992). Beamed Microwave Power Transmission and its Application to Space. *IEEE Transaction on Microwave Theory and Techniques,* vol. 40, no. 6, pp. 1239-1250

Cannon, B. L.; Hoburg, J.F.; Stancil, D.D. & Goldstein, S.C. (July 2009). Magnetic Resonant Coupling As a Potential Means for Wireless Power Transfer to Multiple Small Receivers. *Power Electronics, IEEE Transactions on* , vol.24, no.7, pp.1819-1825

Chen, C.J.; Chu, T.H.; Lin, C.L. & Jou, Z.C. (July 2010). A Study of Loosely Coupled Coils for Wireless Power Transfer. *Circuits and Systems II: Express Briefs, IEEE Transactions on,* vol. 57, no.7, pp.536-540

Greene, C. E; Harrist, D. W.; Shearer, J. G.; Migliuolo, M. & Puschnigg, G. W. (2007). Implementation of an RF Power Transmission and Network. *International Patent Application,* WO/2007/095267

IEEE-SA Standards Board (1999). IEEE Standard for Safety Levels with Respect to Human Exposure to Radio Frequency Electromagnetic Fields, 3 kHz to 300 GHz. *IEEE Std. C95.1*

Imura, T. & Hori, Y. (2011). Maximizing Air Gap and Efficiency of Magnetic Resonant Coupling for Wireless Power Transfer Using Equivalent Circuit and Neumann Formula. *Industrial Electronics, IEEE Transactions on,* vol. 58, no. 10, pp. 4746-4752

Karalis, A.; Joannopoulos, J.D. & Soljacic, M. (January 2008). Efficient Wireless Non-radiative Mid-range Energy Transfer. *Annals of Physics,* vol. 323. No. 1, pp.34-48

Sample, A. P.; Meyer, D.A. & Smith, J.R. (February 2011). Analysis, Experimental Results, and Range Adaptation of Magnetically Coupled Resonators for Wireless Power Transfer. *Industrial Electronics, IEEE Transactions on* , vol.58, no.2, pp.544-554

Tesla, Nikola (1919). The True Wireless. *Electrical Experimenter*, USA

13

A High Frequency AC-AC Converter for Inductive Power Transfer (IPT) Applications

Hao Leo Li, Patrick Aiguo Hu and Grant Covic
The University of Auckland,
New Zealand

1. Introduction

A contactless power transfer system has many advantages over conventional power transmission due to the elimination of direct electrical contacts. With the development of modern technologies, IPT (Inductive Power Transfer) has become a very attractive technology for achieving wireless/contactless power transfer over the past decade and has been successfully employed in many applications, such as materials handling, lighting, transportation, bio-medical implants, etc. (Kissin et al. 2009; Li et al. 2009).

A typical configuration of an IPT system is shown in Fig.1. The system comprises two electrically isolated sections: a stationary primary side, and a movable secondary side.

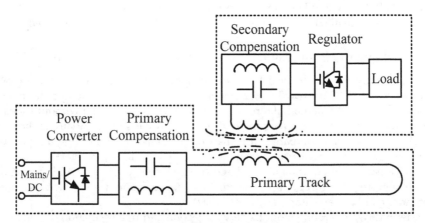

Fig. 1. Typical configuration of a contactless power transfer system.

The stationary primary side is connected to a front-end low frequency power source, which is usually the electric utility at 50Hz or 60Hz, single-phase or three-phase. For some special applications, the power source can be a DC source or a battery. The primary side consists of a high frequency power converter which generates and maintains a constant high frequency AC current in a compensated conductive track loop/coil normally within the range from 10 kHz-1MHz (Dawson et al. 1998; Dissanayake et al. 2008). The pickup coil of the secondary side is magnetically coupled to the primary track to collect energy. The reactance of the secondary side increases proportionally with an increase in operating frequency, and as such is normally compensated by other capacitors or inductors. In order to have a controlled output for different loads, usually a switch mode regulator is used on the secondary side to control the power flow and maintain the output voltage to be constant.

At present, a DC-AC inverter is a common solution to generate a high frequency track current for an IPT system. Often a front-end low frequency mains power source is rectified into a DC power source, and then inverted to the required high frequency AC track current. Energy storage elements, such as DC capacitors, are used to link the rectifier and the inverter. These energy storage elements cause the AC-DC-AC converters to have some obvious drawbacks such as large size, increased system costs, and more complicated dynamic control requirements in practical applications. In addition, those extra components and circuitry reduce the overall efficiency of the primary converter. Having an IPT power supply without energy storage is an intrinsically safe approach for applications and desirable.

Ideally an direct AC-AC converter would be a good alternative to obtain this high frequency power directly from the mains (Kaiming & Lei 2009). A matrix structure eliminates the need for the DC link, but the synchronization between the instantaneous input and output becomes very difficult, and the quality of the output waveform is usually poor due to complicated switching combinations involved (Hisayuki et al. 2005). Furthermore, the circuit transient process involved in the traditional forced switched matrix converters is normally complex and difficult to analyse. The control complications and synchronization limitations make traditional matrix converters unsuitable for IPT systems.

This chapter presents a direct AC-AC converter based on free circuit oscillation and energy injection control for IPT applications. A simple but unique AC-AC topology is developed without a DC link. A variable frequency control and commutation technique is developed and discussed. The detailed circuit model and the converter performance are analysed.

2. Fundamentals of circuit oscillation and energy injection control

Most of the existing converters for IPT applications are resonant converters, where the track is tuned with one or more reactive components in series, parallel or hybrid connection. Regardless of the tuning method, if a resonant tank is oscillatory, even without excitation, a resonant current will oscillate freely provided some energy is stored initially in the resonant tank. A simple free oscillation path can be naturally formed by connecting a capacitor or a tack inductor. This can be achieved in many ways using a switching network.

Fig.2 shows a basic configuration of a voltage sourced energy injection and free oscillation inverter. It comprises a power supply, a switching network and a resonant tank consisting of a track inductor L, a capacitor C and a resistor R.

The inverter has two operating modes: energy injection and free oscillation. When terminals a and b are connected to the power source by the switching network during a suitable period, energy can be injected into the resonant tank. However, when the terminals a and b are shorted by the switching network, the track inductor L, its tuning capacitor C and the resistor R form a free oscillation network, which is decoupled from the power supply. The stored energy in the closed path of a resonant tank will oscillate in the form of an electric field in the capacitor and magnetic field in the inductor, and finally will be consumed by the equivalent resistance which represents the load and the ESR. To maintain the required energy level in the resonant tank for sustained oscillation and energy transfer to any attached loads, more energy is required to go into the tank by reconfiguring the switch network to connect to the power source. From an energy balance point of view, such an operation based on discrete energy injection and free oscillation control is very different from normal voltage or current fed inverters. Therefore, the controller design and performance of the inverters based on this approach are very different from other traditional controllers as well.

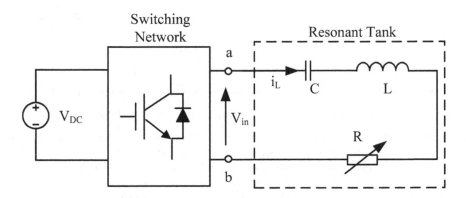

Fig. 2. Principle diagram of injection method of voltage source.

The inverter has two operating modes: energy injection and free oscillation. When terminals a and b are connected to the power source by the switching network during a suitable period, energy can be injected into the resonant tank. However, when the terminals a and b are shorted by the switching network, the track inductor L, its tuning capacitor C and the resistor R form a free oscillation network, which is decoupled from the power supply. The stored energy in the closed path of a resonant tank will oscillate in the form of an electric field in the capacitor and magnetic field in the inductor, and finally will be consumed by the equivalent resistance which represents the load and the ESR. To maintain the required energy level in the resonant tank for sustained oscillation and energy transfer to any attached loads, more energy is required to go into the tank by reconfiguring the switch network to connect to the power source. From an energy balance point of view, such an operation based on discrete energy injection and free oscillation control is very different from normal voltage or current fed inverters. Therefore, the controller design and performance of the inverters based on this approach are very different from other traditional controllers as well.

In principle, any power source may be used to generate high frequency currents apart from a DC source using free oscillation and energy injection control providing the converter topologies are properly designed. For such a reason, if the energy injection control and free oscillation is well coordinated, the energy storage components of an AC-DC-AC converter can be fully eliminated. Therefore, an AC power source can be directly used to generate a high frequency AC current for an IPT system. As a result, the cost, size and efficiency of a primary IPT converter can be significantly improved.

Eliminating the front-end AC-DC rectification and DC storage capacitors, a conceptual AC converter based on energy injection control can be created as shown in Fig. 3.

It can be seen from Fig.3 that an AC source is directly connected to a resonant network by switches. The design of the switching topology could be very critical here to ensure the

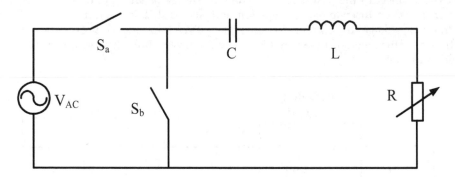

Fig. 3. A conceptual direct AC-AC converter with energy injection control.

energy can be injected according to the load requirements, and to ensure that energy flowing back to the power source is prevented during circuit oscillation.

In practice, most semiconductor switches such as IGBTs and MOSFETs have anti-parallel body diodes. Such a structure ensures the switches can operate bi-directionally but with only one controllable direction. With a combination of the IGBTs or MOSFETs, an AC switch can be constructed to achieve bidirectional controllability (Sugimura et al. 2008). There are many combinations of an AC switches can be used to replace the ideal switches in Fig. 3. After taking the practical consideration of implementation such as control simplicity, cost and efficiency into consideration, the proposed converter topology is developed as shown in Fig. 4, which consists of minimum count of four semiconductors.

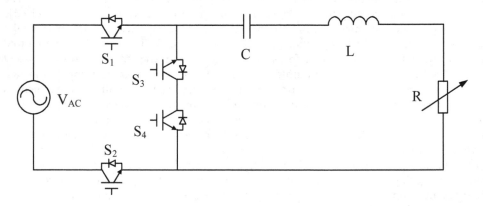

Fig. 4. A typical configuration of a direct AC-AC converter.

The ideal switch Sa in Fig. 3 is presented by an AC switch S1 and S2 as shown in Fig. 4. The ideal switch Sb is replaced by S3 and S4 to construct a free oscillation path for the current. By turning on/off the switches through a properly designed conduction combination, the energy injection and free oscillation can be maintained while the undesired energy circulation to the source can be prevented. The detailed control scheme for all those switches is discussed in the following section.

3. Operation principles

3.1 Normal switching operation

The proposed AC-AC converter is based on a direct conversion topology without a middle DC link. Therefore, the commutation and synchronization of the source voltage and resonant loop branches needs to be considered. To determine the switch operation, it is necessary to identify the polarity of the input voltage and the resonant current. According to the polarity of the low frequency input voltage and the resonant track current, the converter operation can be divided into four different modes as shown in Fig.5.

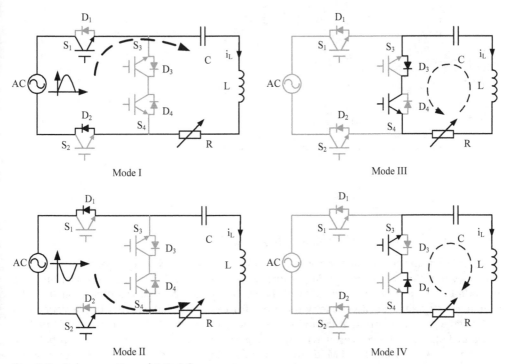

Fig. 5. Switch operation of AC-AC converter.

In Fig.5, Mode III and Mode IV present the current free oscillation in different directions. Mode I and Mode II are the states to control the energy injection based on the different polarities of the input voltage, which is also determined by the directions of the resonant current.

A typical current waveform of the converter and the associated switching signals when the input voltage is in the positive polarity are shown in Fig. 6.

In Fig. 6, if the negative peak value of the resonant current is smaller than the designed reference value -Iref in t1, switch S1 is turned on in the following positive cycle when $V_{AC}>0$, while S2, S3 and S4 remain off. As such, the instantaneous source voltage V_{AC} is added to the resonant tank. This operation results in a boost in the resonant current during t2. Regardless of whether the peak current is smaller or greater than the reference value -Iref, the operation of the converter in the next half cycle of t3 would automatically be switched to Mode III, where switch S3, S4 are on and S1, S2 are turned off, such that the L-C-R forms a

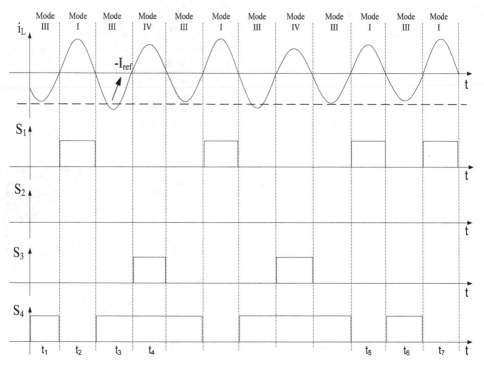

Fig. 6. Converter operation when input voltage $V_{AC}>0$.

free oscillation circuit enabling the energy to circulate between the capacitor C and inductor L. However, if the peak current is larger than the reference -Iref at t3, the converter operates in Mode IV at the next half cycle of t4. If the negative peak current is still smaller than the reference -Iref after a positive energy injection (for example, peak current is still very small at t6 even after the injection at t5, and more energy is still needed in the next positive half cycles of t7.), then, the converter will operate at Mode I at t7, and continue to repeat the operation between Mode I and Mode III in the following half cycles until its peak magnitude larger than the predefined reference value.

The operation of the converter is similar to the situation when the input voltage $V_{AC}<0$. The only difference is that S2 is used to control the negative input voltage which is applied to the resonant tank when the resonant current is negative. Additionally, the peak current would be compared to $+I_{ref}$ to determine if the converter is operated between Mode II/Mode IV, or Mode IV and Mode III. The selection of the mode during normal operation for both positive and negative input is summarized and shown in Fig.7.

3.2 Switching commutation

The proposed converter inherits the simple matrix structure with less switching components and without commutation capacitors. By utilizing the body diodes of each switch, the circuit resonant current can be naturally maintained, even the switches operate at non ZCS condition. Like a DC-AC inverter, variable frequency control can also be employed in the AC-AC converter. Switches being operated under such a condition are switched at the zero

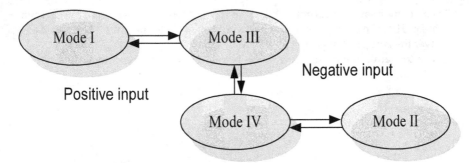

Fig. 7. Mode selection diagram during normal operation.

crossing points to follow the resonance of the current so as to keep the magnitude of the current constant. Nevertheless, the variable frequency switching control also faces problems as a result of the frequency shift (Hu et al. 2000), which cause uncertainty in the direction of the resonant current when the input voltage is at its zero crossing points. This implies that the current completes an entire half cycle over the zero crossing point of input. The polarity change of the input voltage will make the modes of converter operation vary between energy injection states and free oscillation states. Therefore, it is necessary to consider the best switching commutation technique when a variable frequency control strategy is developed for the proposed AC-AC converter.

Theoretically, the track current may have four possible operating conditions around the zero crossing points of the source voltage according to the current directions and the variation tendency of the input voltage, as follows:

Condition A: The input voltage changes from positive to negative. The current stays positive over the zero crossing point as shown in Fig.8.

Here S1 is on to maintain the current while S3 is to be turned on to continue the current.

Condition B: The input voltage changes from positive to negative. The current stays negative over the zero crossing point as shown in Fig.8.

Here S4 is on to maintain the current while S2 is to be turned on to continue the current.

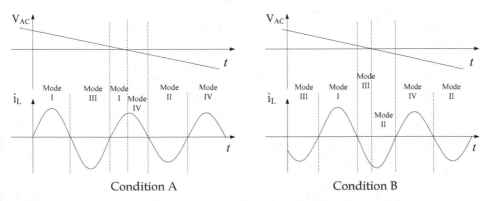

Fig. 8. Operation transient when the input voltage from positive to negative.

Condition C: The input voltage changes from negative to positive. The current stays positive over the zero crossing point as shown in Fig.9.

Here S3 is on to maintain the current while S1 is to be turned on to continue the current.
Condition D: The input voltage changes from negative to positive. The current stays negative over the zero crossing point as shown in Fig.9
Here S2 is on to maintain the current while S4 is to be turned on to continue the current.

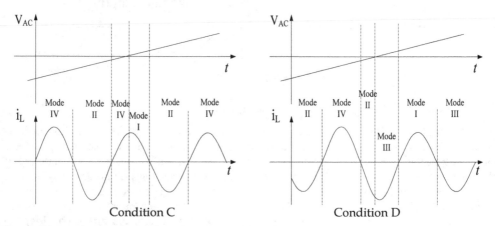

Fig. 9. Operation transient when the input voltage from negative to positive.

It can be seen that there are two additional conditions apart from the normal operation conditions. The operation mode around the zero crossing periods of the input voltage must change between Mode I and Mode IV, or Mode II and Mode III. In fact, the switching commutation between Mode I and Mode IV can be achieved if S3 is always on when the input voltage is in the positive polarities. Similarly, the switching commutation between Mode II and Mode III also can be achieved by keeping S4 on if the input voltage is negative. Such switch operations can maintain the oscillation without affecting normal operation.

From the above discussion, the detailed operation of the AC-AC converter for all conditions is listed in Table 1, according to the input voltage, the resonant current and the predefined current reference. Typical waveforms of the converter with a smooth commutation using this variable frequency control strategy are illustrated in Fig. 10.

The detailed shifting relationship between the operation modes during operation can be summarized in Table 1.

Resonant Current	Input Voltage	Switches/Diodes status	Mode
$i_L > 0$, and previous $\hat{i}_L > -I_{ref}$	$V_{AC} > 0$	S1/D2on S2/S3/S4/D1/D3/D4off	Mode I
$i_L < 0$, and previous $\hat{i}_L < +I_{ref}$	$V_{AC} < 0$	S2/D1 on S1/S3/S4/D1/D3/D4off	Mode II
$i_L < 0$	$V_{AC} > 0$	S4/D3on S1/S2/S3/D1/D2/D4off	Mode III
Previous $\hat{i}_L > +I_{ref}$	$V_{AC} < 0$		
$i_L < 0$	$V_{AC} < 0$	S3/D4on S1/S2/S3/D1/D2/D3off	Mode IV
Previous $\hat{i}_L < -I_{ref}$	$V_{AC} < 0$		

Table 1. Switching states of operation modes.

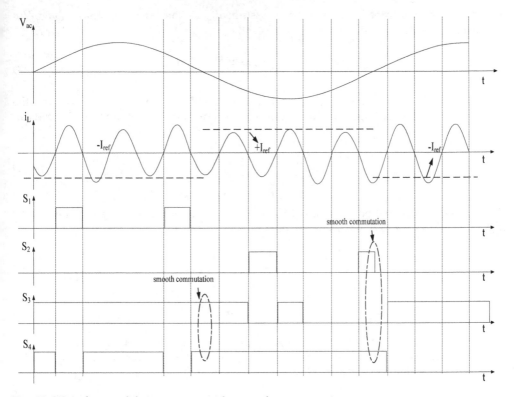

Fig. 10. Waveforms of the converter with smooth commutation.

4. Modeling and analysis

The control strategy of the proposed converter is discrete energy injection control based on the polarity of the input voltage and the resonant current. Since the energy injection occurs discretely, the input phase angle is different for each injection period. During the energy injection period while $V_{AC}>0$, the input voltage is in the same direction as the track current in its positive direction. However, the value of input voltage during each injection varies according to the time instant of the injection. The situation is similar when $V_{AC}<0$; but the track current would be in the negative direction for energy injection. At each injection instant the input voltage over the track can be expressed by the instantaneous value of the AC source as:

$$v_{in}(t) = \hat{V}_{AC} \sin \beta \tag{1}$$

where $\hat{V}_{AC}$ is the peak value of the mains voltage, $\beta=\sin(\omega t)$ is the phase angle of the input AC voltage when energy injection occurs. Theoretically, the instantaneous input voltage during each injection period varies between the beginning and the end of the period. This variation in the input is very small if the resonant frequency is much higher than the mains frequency, as is normally the case for IPT systems. For such a reason, the input voltage applied in each half injection period can be assumed to be constant. The input voltage Vin can therefore be defined as:

$$V_{in} = \begin{cases} \hat{V}_{AC}\sin\beta \\ 0 \end{cases} \tag{2}$$

According to the control strategy of the converter, the differential equations of the equivalent circuit according to the Kirchhoff's voltage law of the circuit can be expressed as:

$$\begin{cases} \dfrac{di_L}{dt} = \dfrac{V_{in}}{L} - \dfrac{Ri_L}{L} - \dfrac{v_C}{L} \\ \dfrac{dv_C}{dt} = \dfrac{i_L}{C} \end{cases} \tag{3}$$

By solving these equations, the instantaneous value of the current during the energy injection and free oscillation periods can be expressed respectively as:

$$i_L = \frac{\hat{V}_{AC}\sin\beta + v_C(0)}{\omega L}e^{-\frac{t}{\tau}}\sin\omega t \tag{4}$$

$$i_L = \frac{v_C(0)}{\omega L}e^{-\frac{t}{\tau}}\sin\omega t \tag{5}$$

where τ=2L/R, ω is the zero phase angle frequency. $v_C(0)$ is the initial voltage at the switching transient. It can be seen that the solution to the track current for a direct AC-AC converter is time dependent during different energy injection periods.

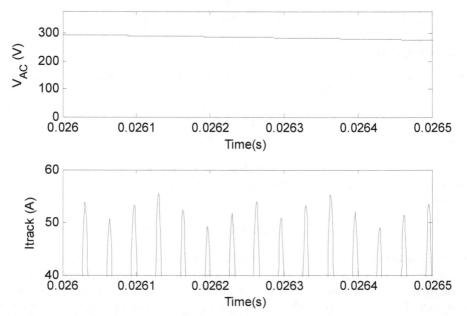

Fig. 11. Current ripples during controlled period.

Under the discrete energy injection control strategy with variable frequency switching, the current resonance can be well maintained, but over energy injection will also occur especially when the AC input voltage reaches its peak. F shows the typical waveform of the track current with a reference of the input voltage.

It can be seen that the current presents a small ripple around the reference current during the controlled period under steady state. The amount of energy injected into the circuit at different phase angle of the mains voltage varies and can be more than what is required to maintain a constant track current. Any over injection of energy during each injection period results in a current overshoot. Similarly, any over consumption during the oscillation period contributes to the current ripple.

An approximated worst case method can be used to find the maximum and minimum peak current of the AC-AC converter caused by the magnitude variation of the AC input. In order to clearly understand the current ripple under the controlled period, a detailed waveform showing the track current under the worst case conditions is given in Fig.12.

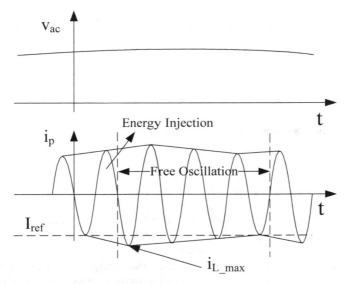

Fig. 12. A detail current ripple waveform.

It can be seen that during the control period, if the input voltage is at the positive cycle, the track current always enters the free oscillation state during its negative cycle. When the converter operates under free oscillation period, there is no energy injected. As a result, the track current is naturally damped by the load and ESR. If the peak current in the negative half cycle is slightly smaller than or equal to the reference value I_{ref}, energy will be injected by the controller into the resonant network in the next positive half cycle. With this energy injection, the current increase from zero to its peak during the positive cycle. Such a peak value will be larger than the previous negative peak value, but it is not the maximum value. This is because the energy will be continually injected during the remaining time of the entire positive cycle and after this positive peak value. In fact, the real maximum peak current i_{L_max} appears in the following negative half cycle as shown in the figure after the half energy injection period is complete.

As stated, in the worst case scenario under no load condition, the capacitor voltage will approximate to zero when the track current is at its peak. After the energy injection over half a period, the total energy storage in the circuit equals the stored energy, and the newly injected energy can be expressed as:

$$\frac{1}{2}L\hat{i}_{L\_max}^{\,2} = \frac{1}{2}LI_{ref}^{\,2} + \int_0^{\frac{T}{2}} v_{in}(t) \cdot i_L(t)dt \tag{6}$$

An energy balance principle can be applied to any energy injection period. But the worst case occurs when the current is just smaller than Iref and energy is still being injected when the mains voltage is at its peak value as stated earlier.

As the frequency of the resonant current is much higher than the 50Hz input voltage, the voltage added to the resonant tank at the peak of 50 Hz can be approximately expressed as:

$$v_{in}(t) \approx \hat{V}_{AC} \tag{7}$$

Therefore the injected energy during the entire half period over the peak of the mains can be expressed by:

$$E_{in} = \int_0^{\frac{T}{2}} v_{in}(t) \cdot i_L(t)dt = \frac{\hat{V}_{AC}^{\,2} + 2\pi f \hat{V}_{AC} I_{ref} L}{2\pi^2 f^2 C} \tag{8}$$

From (6) and (8), the maximum track current can be obtained as:

$$\hat{i}_{L\_max} = \sqrt{I_{ref}^{\,2} + \frac{2T\hat{V}_{AC}\sqrt{LC}(I_{ref}\sqrt{\frac{L}{C}} + \hat{V}_{AC})}{\pi}} \tag{9}$$

It can be seen from equation (9) that the overshoot of the maximum track current is determined by the peak AC input voltage, the controlled reference current, the track current resonant frequency, and the circuit parameters.

Although the maximum peak current is caused by the energy injection, the minimum peak current i_{L_min} is caused by circuit damping. The worst case scenario arises when the load is at its maximum and the peak current is slightly larger than the reference value. Under such a condition there is no injection in the next half cycle. Strictly speaking when the current is at its peak, the capacitor voltage is not exactly zero due to the existence of the load resistance. But for inductive power transfer applications, the Q of the primary circuit is normally high so that the assumption of the initial conditions $i_L(0)= -I_{ref}$ and $v_C(0)=0$ does not cause any significant error. For the proposed AC-AC converter, if it is operated when the input voltage is in its positive cycle, the energy can only be injected in the following positive half cycle. The initial peak value under such a condition is equal to the reference value during negative cycles of the resonant current, and there is no energy injection in the following positive cycle of the current while the damping remains. Instead of damping in the positive half cycle, the damping of current would last for another negative half cycle. With such given initial conditions, the minimum peak current i_{L_min} can therefore be obtained as:

$$\overset{\wedge}{i_{L\_min}} = I_{ref}\frac{\omega_0}{\omega}e^{-\pi R\sqrt{\frac{C}{L}}} \tag{10}$$

According to the structure and control strategy of the AC-AC converter, the minimum current can only happen in the negative cycles. It can be seen from equation (10) that the worst minimum peak current is determined by the reference current and the circuit parameters.

In addition to the current ripples caused by energy injection and energy consumption during the current controlled period, current sag occurs when the input voltage changes its polarity. Fig.13 shows the envelope of the typical current waveform when the input voltage changes from the positive half cycle to the negative half cycle. It can be seen that around the zero crossing point, the input voltage falls back to zero and the magnitude of voltage is very low. Consequently, there is not enough energy that can be injected to sustain the track current to be constant, even if the maximum possible energy is injected in every half cycle.

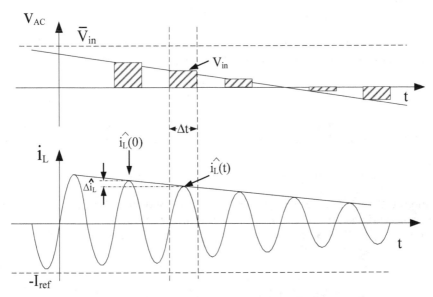

Fig. 13. Current sags around zero crossing point of the input voltage.

It can be seen that before the current sag occurs, the controlled current is around the reference value I_{ref} although small fluctuations exist due to the control. This means the input voltage is large enough to supply the energy to maintain a constant current around I_{ref}. However, over time, the input voltage drops to the boundary value, which is the minimum value to ensure the desired current without any control. This value can be obtained by:

$$\overline{V_{in}} = \frac{RI_{ref}\pi}{2} \tag{11}$$

Theoretically, if the input voltage is smaller than the boundary value, the newly injected energy will be too small to achieve the desired outcome and reaches zero when the input voltage is at the zero crossing point. During this period, the injected energy in each positive half cycle can be calculated by:

$$E_{in} = \int_0^{\Delta t} v_{in}(t) i_L(t) dt = \frac{V_{in} \cdot \hat{i_L} \cdot \Delta t}{\pi} \tag{12}$$

Apart from this newly injected energy in each half cycle, there will be some stored energy in the resonant network because of the energy storage components. As discussed before, because of the high Q characteristic on the primary track, the phase angle of the resonant current and the capacitor voltage is very small and can be ignored. Therefore, the stored energy on the capacitor can be treated as zero when the resonant current is at its peak value. This stored energy in the tank can be expressed in terms of the inductance. At each positive half cycle, the instantaneous stored energy in the resonant tank can be calculated and expressed as:

$$E_{store} = \frac{1}{2} L \hat{i_L}(t)^2 \tag{13}$$

In order to identify the stored energy in different cycles, as shown in Fig. 13, the stored energy during the previous energy injection cycle is equal to:

$$E_{store} = \frac{1}{2} L \hat{i_L}(0)^2 \tag{14}$$

Since the difference in the peak current in two continuous positive half cycle is very small, the variation of the stored energy in one resonant period can be approximately expressed as:

$$\Delta E_{stored} = \frac{1}{2} L (\hat{i_L}(t)^2 - \hat{i_L}(0)^2) \approx L \cdot \Delta \hat{i_L} \cdot \hat{i_L}(t) \tag{15}$$

In addition to the variation in stored energy variation and the injected energy during the zero crossing periods, the energy is consumed by the load during on each half cycle. The consumed energy by the load during each resonant period can be expressed as:

$$E_{consumption} = \frac{\Delta t \cdot \hat{i_L}(t)^2 \cdot R}{2} \tag{16}$$

According to the energy balance principle, the injected energy during each cycle should be equal to the stored and consumed energy when the input voltage is around its zero crossing point. Thus, the equation for expressing the total energy balance in the resonant tank can be obtained as:

$$E_{in} = \Delta E_{store} + E_{consumption} \tag{17}$$

Substituting equation (12) to equation (17), the relationship between the minimum peak current and input voltage can be determined by:

$$\pi L \frac{\Delta \hat{i_L}}{\Delta t} + \frac{\pi}{2} R \hat{i_L} - V_{in} = 0 \tag{18}$$

Because $\Delta \overset{\wedge}{i_L}$ and Δt are very small, equation (18) can be expressed in the format of a differential equation as:

$$L\frac{d\overset{\wedge}{i_L}}{dt} + \frac{R}{2}\overset{\wedge}{i_L} - \frac{1}{\pi}V_{in} = 0 \qquad (19)$$

It can be seen from this equation that the analysis of the peak current of the track is simplified to a first order differential equation with initial values. The envelope can be presented according to the calculated solutions in the time domain.

The input voltage during the current controlled operation period can be modelled as a constant voltage $\overline{V_{in}}$ which is used to maintain the reference current I_{ref}. Therefore, the initial value of the envelope peak current is the reference current, before the voltage drops to zero from the boundary value $\overline{V_{in}}$. Around the zero crossing point of the mains voltage, the input voltage shows a good agreement with a ramp signal shown in Fig.14. It can be seen that the first trace is the approximated ramp input. In comparison, this ramp signal is compared to a 50 Hz input in the second figure. The error between each other is almost imperceptible during the zero crossing point of the input voltage.

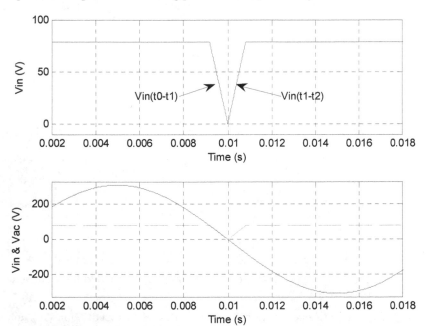

Fig. 14. Piecewise ramp input voltage.

As such, the input voltage of V_{in} around the zero crossing period can be approximated as a piecewise ramp input described by:

$$V_{in}(t_0 - t_1) = \frac{RI_{ref}\pi}{2}\left[1 - \frac{100\pi}{\arcsin(RI_{ref}\pi / 2\overset{\wedge}{V_{AC}})}t\right] \qquad (20)$$

and

$$V_{in}(t_1 - t_2) = \frac{50 RI_{ref}\pi^2}{\arcsin(RI_{ref}\pi / 2\hat{V}_{AC})} t \qquad (21)$$

A figure of the response envelope current with the piecewise input voltage is shown in Fig.15. In order to clearly see the envelope current obtained by equation (19) under the defined voltage of equation (20) and (21), the simulation results of track current by PLECS during the zero crossing point of the mains voltage is shown in the figure also.

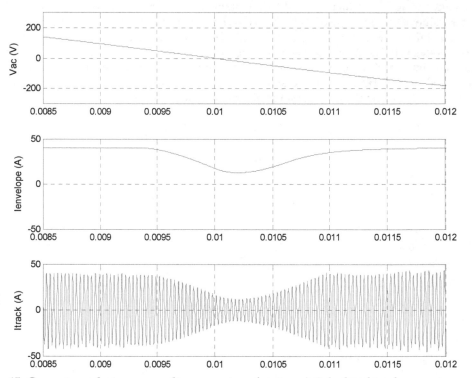

Fig. 15. Current sag during zero voltage crossing of mains: a) Calculated track current envelope, b) Simulated track current.

It can be seen that the envelope of the peak current described by equation (19) under the piecewise input voltage presents a good agreement with the envelope of the simulation track current. If the initial value of the peak current Iref is known, the analytical solution of the minimum value of the current can be obtained by solving equation (19), with the two given piecewise input functions ($V_{in}(t0-t1), V_{in}(t1-t2)$).

By solving equation (19) for the function of the first piecewise input yields:

$$I(t_1) = I_{ref}e^{-\frac{R}{2L}t_1} + K(t_1 - \frac{2L}{R} + \frac{2L}{R}e^{-\frac{R}{2L}t_1}) \qquad (22)$$

Here, time $t_1 = \arcsin(RI_{ref}\pi / \hat{V}_{ac})/100\pi$. K is a constant which can be expressed by $K = -100\pi I_{ref} / \arcsin(RI_{ref}\pi / 2\hat{V}_{ac})$.

After obtaining the final value of the first damping ramp input, the final value of the first piecewise input will be the initial value for the next increasing ramp piecewise input voltage Vin(t1-t2). The solution of the current envelope during this time period of t1-t2 can be expressed by:

$$I(t) = K(t + t_1 e^{-\frac{R}{2L}t}) + \frac{2L}{R}(1 - e^{-\frac{R}{2L}t}) - \frac{2L}{R}e^{-\frac{R}{2L}t}(1 - e^{-\frac{R}{2L}t_1}) + I_{ref}e^{-\frac{R}{2L}(t_1+t)} \quad (t_1 < t < t_2) \qquad (23)$$

It can be seen from Fig.14 that the second piecewise input voltage increases from zero to the reference value $\bar{V}_{in}$ during the time interval t1-t2, the current however reaches a minimum a short period after this. This is because initially the injected energy is very low and the consumed energy is larger than the injected energy. This means that the stored energy is needed to compensate for the over energy consumption. The total energy in the resonant tank would therefore decrease. With the voltage increasing, correspondingly the injected energy would increase. When the injected energy is larger than the energy consumption, the energy starts to increase. As a result the current envelope increases after reaching its minimum value. The exact time can be obtained during t1-t2 by taking:

$$I'(t) = 0 \qquad (24)$$

Therefore the time of the minimum of the current envelope can be obtained as:

$$t_{min} = -\frac{\ln\dfrac{K_1}{\tau K_2}}{\tau} \qquad (25)$$

where $\tau = 2L/R$, $K_1 = K/L$ and $K_2 = 2K/R + I(t_1)$. After knowing the exact time at which the minimum peak current occurs, the minimum value can be obtained by substituting equation (25) into equation (23). The minimum sag current can then be expressed by:

$$I_{min} = \frac{2LK}{R} + \left[I_{ref}e^{-\frac{R}{2L}t_1} + K(t_1 - \frac{2L}{R} + \frac{2L}{R}e^{-\frac{R}{2L}t_1})\right]\frac{2LK_1}{RK_2} \qquad (26)$$

The minimum value of the current sag of the converter is related to the circuit parameters. Ideally, if the consumption is very low and the stored energy is very large, the exponential component has less effect. It requires a large primary Q to act as a filter for filtering off the low frequency components.

5. Simulation study

The simulation study of the AC-AC converter was undertaken using Simulink/PLECS to investigate the operation and analysis of the converter. The PLECS model of the AC-AC converter is shown in Fig.16.

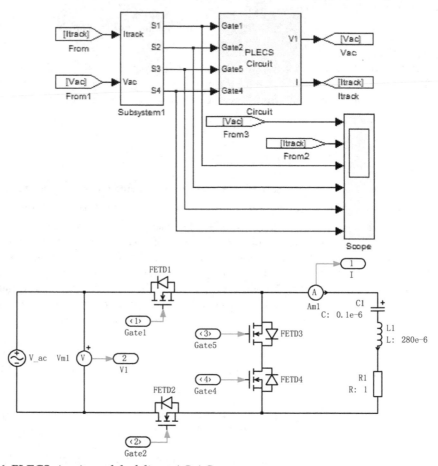

Fig. 16. PLECS circuit model of direct AC-AC converter.

It consists of four switches, an AC input source, and a tuning track. Gate control signals (gate1 to gate4) are fed in from the controller block. The input voltage and track current are measured and feed back to the controller. The converter is designed according to the parameters listed in Table 2.

Symbol	Notes	Value
f_0	Operating frequency of the converter	30 kHz
V_{AC}	Input voltage in RMS	220 V
f_{AC}	Frequency of AC input	50 Hz
L	Track Inductance	280 μH
C	Tuning capacitors	0.1 μF
R	Equivalent total Load	1 Ω
Iref	Track Current reference	40 A

Table 2. Converter circuit parameters of a direct AC-AC resonant converter.

Fig.17. shows the simulation results under variable frequency switching control, which include the waveforms of the input voltage V_{AC}, the track current (I_{track}), and the control signals of S1-S4.

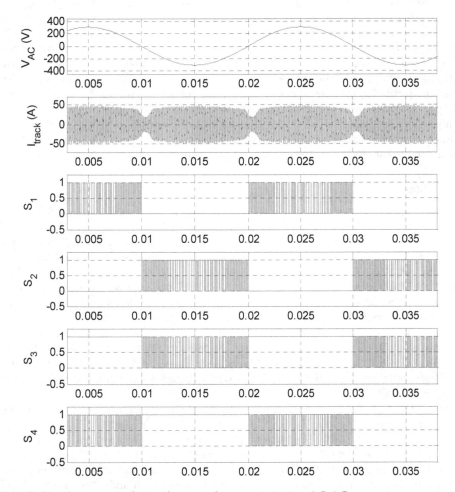

Fig. 17. Simulation waveform of a typical energy injection AC-AC converter.

It can be seen that a low frequency mains voltage can be used to generate a high frequency current for IPT applications under the proposed topology and operation of the AC-AC converter. From Fig. 17 S1 and S4 control the energy injection and the current oscillation when the input is in the positive direction. S2 and S3 control the energy injection and free oscillation during the negative direction of the input voltage. In addition, the switching commutation is achieved smoothly by the proposed switching control technique. The current waveform is controlled around the predefined reference some fluctuations including both the ripples during controlled period and the sages during zero crossing of the input voltage, which have been discussed and compared in the earlier analysis.

6. Conclusions

In this chapter, a direct AC-AC IPT converter has been proposed. The converter has been shown to have a simpler structure compared to a traditional AC-DC-AC converter. This chapter focused on the analysis of the AC-AC converter in relation to the control strategy. While there are a number of possible topologies for the direct AC-AC converters based on energy injection and free oscillation technique as discussed, only one selected example converter topology is described here. The operation principle and a detailed switching control sequence with reference to the current waveforms were analyzed. System modeling and theoretical analysis on the performance of the direct AC-AC converter were also conducted; in particular, the current ripple analysis including the current fluctuation during normal operation was undertaken. In addition, the current sag around zero crossing points of the input voltage was analysed using energy balance principles. In the analysis, the approximate current envelope has also been derived to show the current sag. The validity of both the theoretical analysis and the control method has been verified by simulation studies.

7. References

Dawson, B. V., I. G. C. Robertson, et al. (1998). "Evaluation of Potential Health Effects of 10 kHz Magnetic Fields: A Rodent Reproductive Study."

Dissanayake, T. D., D. Budgett, et al. (2008). Experimental thermal study of a TET system for implantable biomedical devices. IEEE Biomedical Circuits and Systems Conference (BioCAS 2008).

Hisayuki, S., E. Ahmad Mohamad, et al. (2005). High frequency cyclo-converter using one-chip reverse blocking IGBT based bidirectional power switches. Proceedings of the Eighth International Conference on Electrical Machines and Systems.

Hu, A. P., J. T. Boys, et al. (2000). ZVS frequency analysis of a current-fed resonant converter. 7th IEEE International Power Electronics Congress, Acapulco, Mexico.

Kaiming, Y.&L. Lei (2009). Full Bridge-full Wave Mode Three-level AC/AC Converter with High Frequency Link. IEEE Applied Power Electronics Conference and Exposition (APEC 2009).

Kissin, M. L. G., J. T. Boys, et al. (2009). "Interphase Mutual Inductance in Poly-Phase Inductive Power Transfer Systems." IEEE Transactions on Industrial Electronics.

Li, H. L., A. P. Hu, et al. (2009). "Optimal coupling condition of IPT system for achieving maximum power transfer." Electronics Letters 45(1): 76-77.

Sugimura, H., M. Sang-Pil, et al. (2008). Direct AC-AC resonant converter using one-chip reverse blocking IGBT-based bidirectional switches for HF induction heaters. IEEE International Symposium on Industrial Electronics.

Permissions

The contributors of this book come from diverse backgrounds, making this book a truly international effort. This book will bring forth new frontiers with its revolutionizing research information and detailed analysis of the nascent developments around the world.

We would like to thank Ki Young Kim, for lending his expertise to make the book truly unique. He has played a crucial role in the development of this book. Without his invaluable contribution this book wouldn't have been possible. He has made vital efforts to compile up to date information on the varied aspects of this subject to make this book a valuable addition to the collection of many professionals and students.

This book was conceptualized with the vision of imparting up-to-date information and advanced data in this field. To ensure the same, a matchless editorial board was set up. Every individual on the board went through rigorous rounds of assessment to prove their worth. After which they invested a large part of their time researching and compiling the most relevant data for our readers. Conferences and sessions were held from time to time between the editorial board and the contributing authors to present the data in the most comprehensible form. The editorial team has worked tirelessly to provide valuable and valid information to help people across the globe.

Every chapter published in this book has been scrutinized by our experts. Their significance has been extensively debated. The topics covered herein carry significant findings which will fuel the growth of the discipline. They may even be implemented as practical applications or may be referred to as a beginning point for another development. Chapters in this book were first published by InTech; hereby published with permission under the Creative Commons Attribution License or equivalent.

The editorial board has been involved in producing this book since its inception. They have spent rigorous hours researching and exploring the diverse topics which have resulted in the successful publishing of this book. They have passed on their knowledge of decades through this book. To expedite this challenging task, the publisher supported the team at every step. A small team of assistant editors was also appointed to further simplify the editing procedure and attain best results for the readers.

Our editorial team has been hand-picked from every corner of the world. Their multi-ethnicity adds dynamic inputs to the discussions which result in innovative outcomes. These outcomes are then further discussed with the researchers and contributors who give their valuable feedback and opinion regarding the same. The feedback is then collaborated with the researches and they are edited in a comprehensive manner to aid the understanding of the subject.

Apart from the editorial board, the designing team has also invested a significant amount of their time in understanding the subject and creating the most relevant covers. They scrutinized every image to scout for the most suitable representation of the subject and create an appropriate cover for the book.

The publishing team has been involved in this book since its early stages. They were actively engaged in every process, be it collecting the data, connecting with the contributors or procuring relevant information. The team has been an ardent support to the editorial, designing and production team. Their endless efforts to recruit the best for this project, has resulted in the accomplishment of this book. They are a veteran in the field of academics and their pool of knowledge is as vast as their experience in printing. Their expertise and guidance has proved useful at every step. Their uncompromising quality standards have made this book an exceptional effort. Their encouragement from time to time has been an inspiration for everyone.

The publisher and the editorial board hope that this book will prove to be a valuable piece of knowledge for researchers, students, practitioners and scholars across the globe.

List of Contributors

Marco Dionigi and Mauro Mongiardo
University of Perugia, Italy

Alessandra Costanzo
University of Bologna, Italy

Alexey Bodrov and Seung-Ki Sul
IEEE Fellow, Republic of Korea

Héctor Vázquez-Leal, Agustín Gallardo-Del-Angel, Roberto Castañeda-Sheissa and Francisco Javier González-Martínez
University of Veracruz, Electronic Instrumentation and Atmospheric Sciences School, México

Youngjin Park, Jinwook Kim and Kwan-Ho Kim
Korea Electrotechnology Research Institute (KERI) and University of Science & Technology (UST), Republic of Korea

Hiroshi Hirayama
Nagoya Institute of Technology, Japan

Hisayoshi Sugiyama
Dept. of Physical Electronics and Informatics, Osaka City University, Japan

Takashi Komaru, Masayoshi Koizumi, Kimiya Komurasaki, Takayuki Shibata and Kazuhiko Kano
DENSO CORPORATION and the University of Tokyo, Japan

Ick-Jae Yoon and Hao Ling
Dept. of Electrical and Computer Engineering, The University of Texas at Austin, USA

Sungtek Kahng
The University of Incheon, South Korea

Raymond J. Sedwick
University of Maryland, College Park, Maryland, USA

Hunter Hanzhuo Wu
Energy Dynamics Laboratory, Utah State University Research Foundation, USA

Grant Anthony Covic and John Talbot Boys
Department of Electrical and Computer Engineering, The University of Auckland, New Zealand

Huy Hoang and Franklin Bien
Ulsan National Institute of Science and Technology, South Korea

Hao Leo Li, Patrick Aiguo Hu and Grant Covic
The University of Auckland, New Zealand

Printed in the USA
CPSIA information can be obtained
at www.ICGtesting.com
JSHW011453221024
72173JS00005B/1058